Annals of the ICRP

ICRP PUBLICATION 142

Radiological Protection from Naturally Occurring Radioactive Material (NORM) in Industrial Processes

Editor-in-Chief
C.H. CLEMENT

Associate Editor
H. FUJITA

J-F. Lecomte, P. Shaw, A. Liland, M. Markkanen, P. Egidi, S. Andresz, J. Mrdakovic-Popic, F. Liu, D. da Costa Lauria, H.B. Okyar, P.P. Haridasan, S. Mundigl

PUBLISHED FOR

The International Commission on Radiological Protection

by

Please cite this issue as 'ICRP, 2019. Radiological protection from naturally occurring radioactive material (NORM) in industrial processes. ICRP Publication 142. Ann. ICRP 48(4).'

CONTENTS

ICRP Publication 142

Guest Editorial

EXPOSURE TO NATURAL SOURCES OF RADIATION IS NORMAL, BUT EXPOSURE IN THE WORKPLACE SHOULDN'T BE!

If you are reading this editorial, I'm quite confident you woke this morning and didn't give much thought to the multitude of potassium-40 nuclei that decayed in your body, or the secondary cosmic rays that passed through your body during the night. Nor could your human senses detect these radiations or feel the energy deposition in your tissues. Nonetheless, both potassium-40 and cosmic rays will contribute about 0.3 mSv per year to your NORMal radiation dose from background sources. Your cosmic radiation dose will vary somewhat depending on the latitude and altitude of your location, and where we are temporally in the 11-year solar cycle. Additionally, all of us will be exposed to gamma radiation from the natural uranium and thorium series radionuclides in our local soil and rocks, as well as the building materials that compose our homes and places of work. In fact, you may have a granite counter top in your home, or decorative stone in the building where you work, that may contribute to your annual radiation dose. And, all these exposures could be far exceeded by the dose we receive from indoor radon. Exposure to natural sources of radiation is NORMal.

As radiological protection professionals, we understand there is a level of radiation exposure and certain sources we are unable to avoid in daily life, and this has been the situation for millennia. The same will hold true for cosmic radiation exposure in air flight, as we are unable to practically shield such penetrating and complex radiation fields. But we can calculate and track doses for those occupationally exposed. This is particularly important for pregnant air crew. With the large increase in air transportation for the general public during the past 50 years, this would appear to be the new NORMal, and so ICRP has addressed it in *Publication 132* (ICRP, 2016).

In recent years, we have also recognised that radon exposure in homes can be a significant risk factor and cause of lung cancer, and the use of a below-grade soil depressurisation system in our home, school, or workplace building is very effective in mitigating potential radon exposure. ICRP has publications on the risk of lung cancer from radon exposure (ICRP, 2010) and radiological protection measures for radon (ICRP, 2014).

Given the current system of radiological protection developed by ICRP, and implemented by national regulatory bodies, we have 'existing', 'planned', and 'emergency' exposure situations. What I've described above clearly exists for every day of our life, but is unlikely to cause acute tissue damage and warrant an emergency intervention or action. That does not mean there is zero risk from these radiation exposures. The challenge in recent years is the recognition that natural uranium and thorium series radionuclides present in various industrial processes can cause an increased radiation dose to workers and the public, and may result in environmental contamination. Thus, we are now faced with application of the radiological protection principles of 'justification' and 'optimisation' of the exposure, and perhaps application of dose 'limitations' as related to NORM in the workplace. Past industrial operations may have also created legacy contamination that has prompted an evaluation and clean-up. Decommissioning, decontamination, and radioactive waste disposal can be extremely expensive, especially when large volumes of waste are involved. This is often the case with NORM.

This publication provides an excellent framework for dealing with NORM in various industrial processes. Industrial sectors where NORM may become translocated or technologically enhanced to cause worker or public exposure include: mining, metals extraction, water treatment, phosphate, fertiliser, and energy (e.g. coal, oil and gas). Given the relatively low levels of radiation exposure involved and low concentrations of NORM routinely encountered, a 'graded approach' is recommended for the regulatory authorities as well as affected industries. A key aspect for safety professionals and management is an awareness that increased levels of NORM are present. Once realised, it is critical that characterisation of environs, raw materials, products, and waste be performed. This would include external radiation fields and potential for internal intake. From there, qualified personnel can project what worker or public exposure and/or environmental impact may occur. The information contained in this publication provides excellent summaries of typical radiation exposures in various industrial sectors, as well as references to support these data. If radiation exposures or levels of radioactivity demand it, a 'radiological protection management plan' would be advisable. Such a plan may outline appropriate NORM handling; radiation measurement equipment; detailed procedures to perform routine surveys or assays; and how to package, manifest, label, and transport NORM-containing product or waste.

In that NORM materials are not man-made, historically they have not been subject to the controls of reactor- or accelerator-generated material. International and national radiological protection research and regulatory authorities as of late are attempting to bridge that gap. This publication provides a wealth of information that will be useful for these regulatory bodies, as well as the radiological protection practitioner. The authors should be proud of this publication!

References

ICRP, 2010. Lung cancer risk from radon and progeny and statement on radon. ICRP Publication 115. Ann. ICRP 40(1).
ICRP, 2014. Radiological protection against radon exposure. ICRP Publication 126. Ann. ICRP 43(3).
ICRP, 2016. Radiological protection from cosmic radiation in aviation. ICRP Publication 132. Ann. ICRP 45(1).

David J. Allard
Pennsylvania Department of
Environmental Protection

ICRP Publication 142

RADIOLOGICAL PROTECTION FROM NATURALLY OCCURRING RADIOACTIVE MATERIAL (NORM) IN INDUSTRIAL PROCESSES

ICRP PUBLICATION 142

Approved by the Commission in July 2019

Abstract–The purpose of this publication is to provide guidance on radiological protection in industries involving naturally occurring radioactive material (NORM). These industries may give rise to multiple hazards and the radiological hazard is not necessarily dominant. The industries are diverse and may involve exposure of people and the environment where protective actions need to be considered. In some cases, there is a potential for significant routine exposure of workers and members of the public if suitable control measures are not considered. Releases of large volumes of NORM may also result in detrimental effects on the environment from radiological and non-radiological constituents. However, NORM industries present no real prospect of a radiological emergency leading to tissue reactions or immediate danger for life. Radiological protection in industries involving NORM can be appropriately addressed on the basis of the principles of justification of the actions taken and optimisation of protection using reference levels. An integrated and graded approach is recommended for the protection of workers, the public, and the environment, where consideration of non-radiological hazards is integrated with radiological hazards, and the approach to protection is optimised (graded) so that the use of various radiological protection programme elements is consistent with the hazards while not imposing unnecessary burdens. For workers, the approach starts with characterisation of the exposure situation, and integration, as necessary, of specific radiological protective actions to complement the protection strategy already in place or planned to manage other workplace hazards. According to the characteristics of the exposure situation and the magnitude of the hazards, a relevant reference level should be selected and appropriate collective or individual protective actions taken. Exposure to radon is also treated using a graded approach, based first on application of typical radon prevention and mitigation techniques, as described in *Publication 126*. A similar approach should be implemented for public exposure through the control of discharges, wastes, and residues after characterisation of the situation. If the protection of non-human species is warranted, it should be dealt with after an assessment of radiological exposure appropriate for the circumstances, taking into account all hazards and impacts. This should include identification of exposed organisms in the environment, and use relevant derived consideration reference levels to inform decisions on options for control of exposure.

Keywords: Radiological protection; NORM; optimisation; integrated approach; graded approach

MAIN POINTS

- **Exposures resulting from industrial activities involving naturally occurring radioactive material (NORM) are controllable, with protection achieved through justification of taking protective actions and optimisation of protection.**

- **NORM presents no real prospect of a radiological emergency leading to tissue reactions or immediate danger to life; actions to protect workers and the public should consider long-term external exposure, intake of radioactive material, and radon or thoron inhalation.**

- **An integrated and graded approach is recommended for the protection of workers, the public, and the environment, including characterisation of the exposure situation, and optimisation of radiological protective actions to complement the protection strategy already in place or planned to manage other hazards.**

- **Reference levels (excluding exposure to radon and thoron) for the protection of workers should reflect the distribution of exposures and would, in the majority of cases, be less than a few mSv annual effective dose. Very rarely would it be expected that a value exceeding 10 mSv annual effective dose would be necessary.**

- **Reference levels for protection of the public should reflect the distribution of exposures, and would generally be less than a few mSv annual effective dose.**

- **Radon and thoron exposures should be managed using a graded approach, first relying on radon prevention and mitigation measures in the building, as recommended in *Publication 126* (ICRP, 2014b).**

1. INTRODUCTION

1.1. Background

(1) All minerals and raw materials of a geological nature contain radionuclides of natural origin. The main radionuclides of interest are ^{40}K and radionuclides from the ^{232}Th and ^{238}U decay series. ^{232}Th and ^{238}U decay through a series of radionuclides to stable isotopes ^{208}Pb and ^{206}Pb, respectively, known as 'daughter radionuclides' or 'progeny'. The other primordial radionuclides are of much lower abundance or radiological significance.

(2) For most human activities involving minerals and raw materials, the level of exposure due to primordial radionuclide decay series is not a concern for radiological protection. However, there are a number of circumstances in which materials containing naturally occurring radionuclides are recovered, processed, used, or moved such that enhanced radiation exposures may result. Material involved in processes giving rise to these enhanced exposures is considered to be naturally occurring radioactive material (NORM). For example, certain minerals, including some that are commercially exploited, may contain potassium and/or thorium and/or uranium progeny at significant concentrations.

(3) Furthermore, during the extraction of minerals and their processing, the radionuclides may be dispersed and/or their physicochemical properties changed so that they become unevenly distributed between the products, by-products, discharges, residues, or wastes arising from the process(es). The radionuclide activity concentrations may exceed those in the original mineral, sometimes by several orders of magnitude, which can significantly increase the exposure of workers and/or members of the public, and lead to contamination of the environment.

(4) Only 2 years after the discovery of radioactivity by Becquerel in 1896, Marie Curie identified radium and polonium after processing several tons of pitchblende, an ore with high uranium content. Radon – or 'radium emanation' as it was called – was found a few years later in petroleum and natural gas brought to the surface. Later, several investigations led to the first general review of radioactivity associated with sedimentary rocks, petroleum, underground water, and brines (Monicard and Dumas, 1952). The discovery of radioactive scale from natural sources, for instance in British and American oil production facilities, was first mentioned in the 1950s (Schmidt, 2000). However, the potential health, safety, and environmental risks due to NORM exposure in the industry have only been widely recognised since the 1980s (Miller et al., 1991).

(5) In *Publication 26* (ICRP, 1977), the Commission recognised that some practices may 'increase the level of exposure from the natural background of radiation' (Para. 235), and that there may be levels of natural radiation that might have to be controlled in much the same way as for artificial sources. The Commission did not give practical guidance on the principles for such control. In the same year, the United Nations Scientific Committee on the Effects of Atomic Radiation (UNSCEAR) introduced a chapter on 'technologically enhanced exposures to natural radiation' in its report to the General Assembly (UNSCEAR, 1977).

(6) In *Publication 39* (ICRP, 1984) and later in *Publication 60* (ICRP, 1991), the Commission proposed principles for limiting exposure of workers and the public to natural sources of radiation, notably primordial radionuclides and progeny. The Commission stated that there should be requirements to include some exposures to natural sources as part of occupational exposures when it comes to 'operations with and storage of materials not usually regarded as radioactive, but which contain significant traces of natural radionuclides' (ICRP, 1991, Para. 136).

(7) In *Publication 82* (ICRP, 1999) devoted to protection of the public against prolonged exposures, the Commission first acknowledged the term 'NORM' by noting: 'industrial development has further increased the "natural" exposure of people by technologically enhancing the concentrations of radionuclides in naturally occurring radioactive materials (NORMs)' (Para. 6). *Publication 82* then focused on application of the system described in *Publication 60* (ICRP, 1991) for radiological protection to practices resulting in prolonged exposure. Optimisation was expected to be applied to ensure that doses were 'as low as reasonably achievable', taking into account economic and social factors. The Commission later provided detailed guidance on application of the optimisation principle in *Publication 101* (ICRP, 2006). This publication recommended that dose constraints and dose limits for practices may be appropriate to NORM exposure, but should be applied with 'care and flexibility'.

(8) In *Publication 103* (ICRP, 2007a), the Commission revised the system for radiological protection of *Publication 60* (ICRP, 1991). The approach is now based on the characteristics of the radiation exposure situation, rather than the process-based approach adopted previously. The system applies to all controllable exposures to ionising radiation from any source, regardless of size or origin, but applies in its entirety only to situations in which either the source of the exposure or the pathways leading to doses received by individuals can be controlled by some reasonable means. A major implication of this revision is that all exposures, including those from NORM, are now within the scope of the system, and the principles of justification and optimisation always apply. *Publication 103* (ICRP, 2007a, Para. 284) noted human and environmental exposures resulting from many industries involving NORM as examples of existing exposure situations.

(9) *Publication 104* (ICRP, 2007b) recognised that there is a need for international consensus on NORM exposure management, and that industries involving NORM have been regulated variably with regard to radiological protection because the radiological protection system was introduced after the start of operation, and existing industrial hygiene controls already limit the potential for radiation exposure (e.g. control of airborne dust). Exclusion and exemption of industries involving NORM and activities using numerical criteria may be useful but lack the qualitative judgement that is also often necessary. Hence, *Publication 104* advocated a graded approach in the management of NORM exposure, taking into account the prevailing circumstances and risk to people, with the global aim of promoting the protection of workers and public health (Para. 137).

(10) Following the 2007 Recommendations (ICRP, 2007a), several publications are intended to cover protection of the environment more explicitly, particularly

Publication 108 (ICRP, 2008), which introduces the Reference Animals and Plants (RAPs), and *Publication 124* (ICRP, 2014a), which deals with the application of protection in the environment under different exposure situations.

(11) The Commission has recently engaged in a set of publications dedicated to applying the system of radiological protection to existing exposure situations. *Publication 126* (ICRP, 2014b) updated the recommendations for protection against exposure to radon. *Publication 132* (ICRP, 2016) is devoted to radiological protection from cosmic radiation in aviation. *Publications 109* and *111* (ICRP, 2009a,b), on emergency exposure situations and living in long-term contaminated areas following a radiological emergency, are currently being updated. A publication is also in preparation dedicated to exposures resulting from contaminated sites from past industrial, military, and nuclear activities.

1.2. Scope

(12) The Commission's approach to NORM builds on *Publication 103* (2007 Recommendations), *Publication 104* (scope of the system), *Publication 124* (protection of the environment), and *Publication 126* (protection against radon and thoron) (ICRP, 2007a,b, 2014a,b). The focus is upon industrial processes such as mining and mineral extraction, or other industrial activities that may lead to exposures to NORM of geological origin which have been identified as requiring consideration of radiological protection. The term 'industrial' also includes small-size business activities. In many cases, the raw material does not have elevated levels of NORM (e.g. fossil fuels); however, the subsequent activities and industrial processes generate higher concentrations of radionuclides in the products, by-products, discharges, residues, or wastes. The industrial processes may also increase the exposure of workers and/or members of the public, and/or lead to discharges of radioactive substances to the environment with subsequent human and non-human exposure. More details about activities that may involve NORM exposure are given in Section 2 and Annex A.

(13) This publication outlines how exposures resulting from NORM industries can be managed through justification of the actions taken, optimisation of protection, and use of appropriate individual dose criteria. It does not cover industries or other dealings with NORM where exposures are more appropriately managed as planned exposure situations. This includes, but is not necessarily limited to, mining facilities that have been established for the purpose of extracting materials such as uranium and thorium from ore to be used for their radioactive, fissile, or fertile properties.

(14) One contributor to NORM exposures is radon (^{222}Rn) gas (from the decay of ^{238}U) and, to a lesser extent, thoron (^{220}Rn) gas (from the decay of ^{232}Th). ICRP recently provided information on the risk of lung cancer from radon and thoron by reviewing epidemiological studies (ICRP, 2010), formulated recommendations for protection of workers and the public (ICRP, 2014b), and provided new dose coefficients for radon (ICRP, 2017b). In *Publication 126* (ICRP, 2014b), the Commission

recommends an integrated approach for controlling radon exposure, relying as far as possible on the management of buildings or locations in which radon exposure occurs, whatever the use of the building. This approach is valid for radon and thoron arising from different sources in the workplace (e.g. from the ground, building materials, and minerals containing NORM). Thus, radon and thoron exposures in industries involving NORM should be managed in accordance with the approach of *Publication 126*.

(15) Due to the long-standing history of many industries involving NORM, sites have been identified as contaminated by NORM residues and wastes from past activities (legacy sites). In 2014, ICRP established a task group to develop a report on how to apply the Commission's recommendations to exposures resulting from contaminated sites from past industrial activities, so this topic will not be fully addressed here.

(16) The 2007 Recommendations (ICRP, 2007a) extended the system of radiological protection to address protection of the environment, including flora and fauna, more explicitly. Later, in *Publication 108* (ICRP, 2008), the Commission describes its framework for protection of the environment through the introduction of RAPs, and how it should be applied within the system of radiological protection. *Publication 124* (ICRP, 2014a) deals with the application of protection of the environment under different exposure situations. Consistent with the approach established by the 2007 Recommendations (ICRP, 2007a), the present publication will specifically address protection of the environment against NORM exposure. In complement, *Publication 114* (ICRP, 2009c) proposes transfer parameters for RAPs, *Publication 136* (ICRP, 2017a) recommends dose coefficients for non-human biota, and a report related to the derivation of radiation-weighting factors for application in dose assessment for RAPs is in preparation.

(17) The ethical underpinnings of the system of radiological protection rely on four core ethical values as described in *Publication 138* (ICRP, 2018): beneficence/non-maleficence, prudence, justice, and dignity. There are important ethical issues to be integrated in the protection strategy against NORM exposure. Applying the system of radiological protection is a permanent quest for decisions that do more good than harm (beneficence/non-maleficence), avoid unnecessary risk (prudence), establish a fair distribution of exposures (justice), and treat people with respect (dignity).

(18) While ionising radiation may be a consideration in terms of the protection of people and the environment from NORM, it is generally not the only hazard and perhaps not even the most dominant hazard. Indeed, NORM residues and wastes may contain toxic non-radiological constituents that may be harmful to human health and/or the environment (e.g. heavy metals). The present publication will not provide guidance on the management of these constituents, which may have to be controlled by industrial hygiene and environmental regulation. However, the Commission recommends the use of an integrated approach for the management of radiation and all other hazards that may be present, so that protection is optimised for all concerns in an inclusive manner.

(19) The recommendations in the present publication for radiological protection in industries involving NORM supersede all previous related recommendations in *Publications 103, 104, 124,* and *126* (ICRP, 2007a,b, 2014a,b).

1.3. Structure of this publication

(20) Section 2 presents the characteristics of NORM exposure, an overview of the industries and practices where NORM exposure can occur, and elements related to the NORM cycle. Section 3 describes the Commission's system of radiological protection applied to NORM exposure, including the type of exposure situation, the category of exposure concerned, and the basic principles to be applied. Section 4 provides guidance on implementation of the system of radiological protection using an integrated and graded approach for the various exposed workers, the public, and the environment. Conclusions are provided in Section 5. Annex A provides more details about activities that may involve NORM exposure.

2. CHARACTERISTICS OF EXPOSURE TO NORM

2.1. Ubiquity and variability

(21) Radionuclides of natural origin are ubiquitous and are present in almost all materials on earth. They are generally not of radiological concern. Some human activities, however, have the potential to enhance radiation exposures from these materials.

(22) Many organisations have produced comprehensive reviews of industries that may cause NORM-related radiation exposure of workers, the public, and the environment (UNSCEAR, 1982, 2008; EC, 1999a; IAEA, 2006; EURATOM, 2013). Examples are given below. Further, previous industrial sites could have involved NORM, and these legacy sites may require attention. Details on these work activities are provided in Annex A.

- Extraction of rare earth elements.
- Production and use of metallic thorium and its compounds (i.e. for their metallic, not fissile or fertile, radioactive properties).
- Mining and processing of ores (other than uranium or thorium for the nuclear fuel cycle).
- Oil and gas recovery process.
- Manufacture of titanium dioxide pigments.
- The phosphate mining and processing industry.
- The zircon and zirconia industries.
- Production of metal (tin, copper, iron, steel, aluminium, niobium/tantalum, bismuth, etc.).
- Combustion of fossil fuel (mainly coal).
- Water treatment.
- Geothermal energy production.
- Cement production and maintenance of clinker ovens.
- Building materials (including building materials manufactured from residues or by-products).

(23) Typical industries involving NORM process a wide range of raw materials with different levels of activity concentrations, producing a variety of products, by-products, discharges, residues, and wastes. These industries may or may not be of radiological concern depending on the activity concentrations in the raw materials handled, the processes adopted, the uses of final products, the reuse and recycling of residues, and the disposal of wastes.

(24) The range of processes leads to a large spectrum of scenarios for radiation exposure in workplaces, including:

- large quantities of material as an ore or a stockpile of raw material, residues, or wastes;

- small quantities of material with concentrated radionuclides, such as mineral concentrates, scale, and sludge; and
- material that has been volatilised in high-temperature processes, like slag, precipitator dust, and furnace fumes.

(25) Work activities involving NORM can give rise to external and internal radiation exposures. External exposures can arise from extended exposures to low (gamma) dose rates; shorter exposures to high (gamma and sometimes beta) dose rates from performing maintenance on internals of equipment, slag, scale, and sludge; or a combination of these. The potential for internal exposure is governed mostly by the way NORM appears in the workplace, and the personal protective equipment (PPE) worn by workers. Radon may be an important source of exposure in indoor or underground atmospheres. Indoor radon exposure may arise from the soil, the processed NORM, or the building materials of the facility. In large-scale mining and milling operations, airborne dust is a common industrial hazard, and internal exposures from inhalation of NORM can be significant, especially where higher activity concentrations are present (e.g. above tens of $Bq\,g^{-1}$). In contrast, internal exposures from ingestion of NORM, including in water, are usually low (EC, 1999a). However, there can be considerable differences depending on workplace conditions, the radionuclides involved, and the physical and chemical matrices in which the radionuclides are incorporated (UNSCEAR, 2016).

(26) Very large numbers of workers in the world may be exposed to NORM, and although the data are more limited than those for occupational exposures to man-made sources, the worldwide level of exposure for workers exposed to natural sources of radiation has been estimated at 30,000 man.Sv annually (approximately 13 million workers) (UNSCEAR, 2008). Until implementation of the International Basic Safety Standards for protection against ionising radiation in 1996 (IAEA, 1996), most countries had not been particularly concerned with assessing occupational exposure to natural sources of radiation. Table 2.1 [adapted from IAEA (2006)] gives ranges of exposures to workers in some industries involving NORM. In the majority of workplaces, both the average and the maximum assessed doses received by workers are below a few mSv per year, but higher doses – in some cases, as high as a few tens of mSv – may occur in specific workplaces (approximately $100\,mSv\,year^{-1}$ in very few underground mines).

(27) In terms of public exposure, direct external exposures (i.e. from NORM on the site) are usually negligible, although there are exceptions to this. For some specific industries involving NORM sites, it has been reported that some representative individuals in close proximity to the plant can receive annual doses in the mSv range (UNSCEAR, 2008). In general, public doses from NORM mainly arise from radionuclides released into air and water as routine discharges, and the use of NORM-containing by-products in commodities such as building materials. In rare cases, NORM in drinking water may be an issue. A complete review is made difficult by the diversity of industries involved, the local circumstances associated with exposures, and the overall lack of site-specific radiological assessments. Table 2.2 presents some data

Table 2.1. Examples of dose assessments for workers (external and internal from dust, excluding exposure to radon).

Activities	Radionuclides with highest activity concentrations	Annual effective dose (mSv)			
		Minimum	Mean	Maximum	Distribution
Processing of thorium concentrate[*]	^{232}Th (in feedstock and product)	3.0		7.8	
Production of thorium compounds[†]				82	67% <1
Mining of rare earth ore[‡]	^{238}U and ^{232}Th series (feedstock)		0.24–1		
Beneficiation of rare earth ore[‡]			0.28–0.61		
Handling of monazite	^{232}Th series			0.3	
Rare earth separation and purification	^{228}Ra (residues)			0.3	
Decommissioning of rare earth plant[§]	^{228}Ra (residues)	0.2	7.2	8.94	
Mining of ore other than uranium ore	^{238}U and ^{232}Th series (in general)	1.3	3	5	
Oil and gas production, offshore	^{226}Ra (scale/sludge)			0.5	
Oil and gas production, onshore				0.05	
Oil production, cleaning of pipes[‡,¶]			0.6	3	80% <1
Titanium dioxide pigment production	• ^{232}Th (feedstock) • ^{226}Ra, ^{228}Ra (scale)			0.27	
Phosphate ore storage	^{238}U series			0.28	
Phosphate fertiliser production	• ^{238}U (feedstock and product) • ^{226}Ra (residues)			0.5	
Zircon production	• ^{238}U series (feedstock)			0.4	
Bastnäsite (zirconia) production	• ^{210}Po (in dust precipitator)			0.4	
Manufacture and use of zircon	• ^{238}U (in fused zirconia/ product)	0		2.3	87% <1
Manufacture and use of refractory ceramics		~0.01		1.5	98% <1
Manufacture of zircon/zirconia ceramics		.... Negligible			

(continued on next page)

Table 2.1. *(continued)*

Activities	Radionuclides with highest activity concentrations	Annual effective dose (mSv)			
		Minimum	Mean	Maximum	Distribution
Processing of tin, aluminium, titanium, and niobium ores	• ^{232}Th (feedstock, product and slag) • ^{228}Ra (residue)	0		3.2**	69% <1
Copper smelting	^{226}Ra (slag)			<1	
Recycling of metal scrap	^{210}Po, ^{210}Pb (precipitator dust)	....	Negligible	....	
Coal mining	• ^{238}U • ^{226}Ra, ^{228}Ra (for coal with high Ra inflow water)		2.75		
Combustion of coal	^{210}Po (scale)	0		0.4	
Combustion of coal				<1	
Combustion of coal				0.13	
Drinking water treatment	^{226}Ra (sludge)			<1	
Manufacture of mineral insulation††	NA	0.0011		0.0173	

NA, not available.
*Doses include contributions from inhalation of thoron.
†Doses >1 mSv year^{-1}, mainly due to dust inhalation, were identified in two of the six workplaces investigated. The assessment is being repeated after the implementation of dose reduction measures (equipping workers with respiratory protection, cleaning the workplaces periodically, and installing air filtration).
‡Dose from external exposure alone.
§Doses received over a 9-month decommissioning period.
¶Doses received over a 5-month refurbishment period.
**The maximum dose was 6 mSv prior to 2008.
††The minerals were coal, bauxite, basalt, and cement.

Table 2.2. Examples of dose assessments for members of the public (excluding exposure to radon).

Activities	Radionuclides with highest activity concentrations	Annual effective dose (mSv)
Mining of rare earth ore	^{232}Th (contaminated soil)	0.044
Beneficiation of rare earth ore	^{232}Th (contaminated soil)	0.043
Use of slag from rare earths and steel production in house bricks	^{226}Ra, ^{232}Th (bricks)	~0.2
Production of thorium welding rods	NA	Negligible
Mining of ore other than uranium ore		Specified only as <1
Large mineral residue deposit, 1 Bq g^{-1} ^{238}U and/or ^{232}Th	^{232}Th and ^{238}U series	0.05–0.26
Oil and gas production	NA	Specified only as <1
Elemental phosphorus production		<0.04
Use of dicalcium phosphate animal feed	^{210}Po, ^{210}Pb (in chicken)	<0.02
Use of PG for agriculture	^{226}Ra (in fertiliser)	Negligible
Use of PG for construction of houses	^{226}Ra (in building material)	
Walls and ceilings, PG panels		0.02–0.2
Walls, ceilings, and floor, hollow PG panels		0.46
Walls, ceilings, and floor, solid PG panels		4.5
Walls, PG plasterboard lining		0.15 (India) or insignificant (Australia)
Walls, PG in bricks and cement		≤1.4
Manufacture of zircon/zirconia ceramics		Negligible
Steel production	^{232}Th, ^{228}Ra (in dust/gaseous effluent)	<0.01
Use of metal recycling slag for road construction	^{226}Ra (slag)	Specified only as <1
Combustion of coal	NA	Negligible
Drinking water treatment	NA	Negligible

(continued on next page)

Table 2.2. *(continued)*

Activities	Radionuclides with highest activity concentrations	Annual effective dose (mSv)
Disposal of water treatment residues in landfill	^{226}Ra (sludge)	0.01
Effluent water treatment, former uranium mine	NA	Specified only as <1
Use of common building materials for house construction	NA	<0.3–1

PG, phosphogypsum; NA, not available.

related to public exposures from NORM [adapted from IAEA (2010)]. These estimates are subject to uncertainties and are often conservative. In Table 2.2, the annual effective dose from NORM to the public is estimated to be well below 1 mSv year^{-1}, except in rare cases such as the wide use of phosphogypsum in building material.

(28) *Publication 103* (ICRP, 2007a) introduced an approach for developing a framework to demonstrate radiological protection of the environment. However, to date, there are few examples of assessment of the impact of NORM, outside of uranium mining activities (or similar), on the environment. Each case should be evaluated on an individual basis using an integrated and graded approach, taking all the present hazards, concerned species, main environmental conditions, and other characteristics into consideration.

2.2. A cradle-to-grave perspective on NORM

(29) Several stages of production involving NORM can be identified. Some industries may involve almost all of these stages, whereas others may only involve some of them:

- mineral extraction;
- mineral beneficiation and processing;
- fabrication of products;
- use of products and by-products;
- reuse and recycling of residues;
- management of wastes; and
- dismantling or remediation and rehabilitation.

(30) The presence of NORM with elevated radionuclide concentrations could be an issue at any stage, and may lead to significant radiological exposures of workers

and the public, and contamination of the environment and subsequent fauna and flora exposure if not adequately controlled.

(31) By-products and residues from an industry involving NORM can be used as feedstock by other industries involving NORM and/or in common practice (e.g. building materials). In such circumstances, after being brought to the surface (or introduced into the industrial sector by another means), NORM enters a cycle which is possibly endless (i.e. NORM can be moved and/or reprocessed from place to place). Therefore, the enhanced exposures due to NORM may occur during all stages of the cycle.

3. APPLICATION OF THE COMMISSION'S SYSTEM OF RADIOLOGICAL PROTECTION TO NORM

3.1. Types of exposure situations and categories of exposure

3.1.1. Types of exposure situations

(32) The Commission defines an exposure situation as a 'network of events and situations' that begins with a natural or artificial radiation source, the transfer of the radiation or radioactive materials through various pathways, and the resulting exposure of individuals or the environment (ICRP, 2007a, Para. 169). Protection can be achieved by taking action at the source, or at any point in the exposure pathways of the exposed individuals.

(33) According to Para. 176 of *Publication 103* (ICRP, 2007a), the Commission intends its Recommendations to be applied to all sources and to individuals exposed to radiation in the following three types of exposure situations, which address all conceivable circumstances: existing exposure situations, planned exposure situations, and emergency exposure situations.

(34) The health objectives of the Commission's system of radiological protection are to manage and control exposures so that deterministic effects are prevented, and the risk of stochastic effects is reduced to the extent reasonably achievable (ICRP, 2007a, Para. 29). The situation-based approach is designed in order to frame a graded approach proportionate to the expected level of risk, and notably, the likelihood of deterministic effects. Emergency exposure situations are situations in which urgent protective actions are needed to avoid deterministic effects. Many planned exposure situations also present the possibility of deterministic effects if not controlled properly. This is mainly a consequence of the use of radionuclides for their radioactive, fissile, or fertile properties, or of apparatus that generate radiation such as x rays. On the other hand, existing exposure situations, unlike emergencies, do not require urgent action because the types, forms, and concentrations of radionuclides realistically do not have prospect to cause deterministic effects over a short period of time.

(35) The Commission has considered exposures resulting from many industries involving NORM as examples of existing exposure situations (ICRP, 2007, Paras 284 and 288). For most NORM industries, the source is not deliberately introduced in the industrial process for its radioactive properties. There is generally no prospect of a radiological emergency or deterministic effects. The process by which NORM in raw materials is concentrated, with changes in physicochemical form resulting in production of radioactive release, residues, and wastes, is not for the purpose of introducing a new radioactive source; it is incidental although it has to be managed. However, when NORM is processed for its radioactive, fissile, or fertile properties, the Commission considers it a planned exposure situation.

(36) A distinct feature of industries involving NORM is the fact that the source is modified from its original state, sometimes deliberately. Due to this feature,

regulators are inclined to implement the same types of regulatory processes that are used for man-made sources, particularly for the protection of workers (EURATOM, 2013; IAEA, 2014). This may be as a result of recognising existing activities that may have significant radiological hazards that require an ongoing radiological protection programme. Likewise, a new activity, if recognised to have radiological implications, may suggest the need to use the same tools that are typically associated with radiological protection programmes and regulatory structures for man-made sources posing similar risks. Nonetheless, this feature does not change the relevance of the integrated and graded approach as recommended in this publication.

(37) As mentioned above, the philosophy of *Publication 103* (ICRP, 2007a) compared with *Publication 60* (ICRP, 1991) is to recommend a consistent approach for the management of all types of exposure situations. This approach is mainly based on application of the principle of optimisation using appropriate dose criteria. In existing exposure situations, the relevant dose criterion is the reference level. However, the use of regulatory tools commonly associated with authorised planned exposure situations, such as a dose limit, can be suitable in existing exposure situations when necessary to properly demonstrate ongoing control of the radiological hazards. This application of regulatory tools does not change the features and characteristics of the exposure situation, but may, for convenience, change the regulatory designation. Exposure situations are helpful for considering the relationship between sources and exposures and the corresponding implementation of the radiological protection principles, but the Commission recommends flexibility in the use of regulatory tools to effectively achieve protection. For the protection of non-human species, the use of environmental reference levels based on derived consideration reference levels (DCRLs) is also recommended (ICRP, 2014a). Whatever the type of exposure situation, the aim is to achieve a standard of protection that is proportionate to the expected level of risk.

(38) A graded approach, commensurate to the level of risk as well as other considerations such as economic and societal, is appropriate and particularly relevant for industries involving NORM due to the economic importance of industries, large volumes of residues and wastes, limited options for management, moderate level of doses, and the potentially high cost of regulation in relation to reduction in exposure. Industries involving NORM are generally situations where multiple hazards and pollutants can be present. The radiological risk may not be the dominant hazard, and consequently, there has often been no or only limited radiological protection awareness. In such a context, the radiological protection system is not necessarily the only driving force in safety, and an integrated approach to all hazards should be employed. The graded approach to protection should first take account of the existing knowledge and experience of these industries in the management of industrial hazards, and then pragmatically integrate any additional measures necessary for the purposes of radiological protection.

(39) The doses resulting from the process in which NORM is concentrated are expected to remain relatively low whatever the circumstances. In the same way, the imaginable scenarios of loss of control of the radioactive material in industries

involving NORM result in a limited impact in terms of doses and subsequent health effects such as tissue reaction or immediate danger to life. Consequently, industries involving NORM present no real prospect of a radiological emergency, and thus are not likely to give rise to an emergency exposure situation, but releases and discharges may result in environmental and public exposure.

3.1.2. Categories of exposure

(40) The Commission distinguishes between three categories of exposure: occupational, medical, and public exposures. Occupational exposure is radiation exposure of workers incurred as a result of their work. However, because of the ubiquity of radiation, the Commission traditionally limits the definition of 'occupational exposures' to radiation exposures incurred at work as a result of situations that can reasonably be regarded as being the responsibility of the operating management. Medical exposure is the exposure of patients in the course of medical diagnosis and treatment. Public exposure encompasses all exposures to humans other than occupational exposures and medical exposures of patients.

(41) Industries involving NORM can give rise to both occupational and public exposure, but not medical exposure.

(42) In most cases, the exposure of workers in industries involving NORM is adventitious because the presence of NORM in the material processed is a natural fact, without intentional addition for its radioactive properties, and the workers are often not considered to be occupationally exposed. As indicated in *Publication 126* (ICRP, 2014b, Para. 59), referring to *Publication 65* (ICRP, 1993), workers who are not regarded as being occupationally exposed to radiation are usually treated in the same way as members of the public. However, their exposures should be considered. It is the responsibility of the operating management to integrate the radiation risk with the other hazards, and to address all the hazards in accordance with the agreed standards for health and safety at work.

(43) As described in Section 4.1, a graded approach is recommended for the protection of workers in industries involving NORM, based on selection of the reference level as well as selection and implementation of reasonable protective actions. This approach should also consider, as explained above, the integration of radiological protection in procedures for the control of other hazards in a more global and synergistic approach to hazard management.

(44) In rare cases, the level of dose remains high or the application of special working procedures is needed for radiological protection purposes. In these cases, the measures recommended for occupationally exposed workers would apply (ICRP, 1997). The Commission's recommendations should not be interpreted as requiring all elements of a protection programme irrespective of the circumstances. The approach should be graded based on the hazards present.

(45) Public exposure is addressed through the control of NORM discharges, wastes, residues (including recycling and reuse), and possible legacy sites, as explained in Section 4.2.

(46) Industries involving NORM also generate exposure to the environment through extraction, transportation, storage, processing, effluents, discharges, and accidental releases. As indicated in Section 4.3, exposure to the environment is also dealt with in a graded approach on the basis of common environmental standards and considering the presence of NORM.

3.2. Justification of protection strategies

(47) The principle of justification is one of the two fundamental source-related principles that apply to all exposure situations. The recommendation in Para. 203 of *Publication 103* (ICRP, 2007a) requires, through the principle of justification, that any decision that alters the radiation exposure situation should do more good than harm. The Commission goes on to emphasise that for existing exposure situations, the justification principle is applied when deciding whether to take action to reduce exposure and avert further additional exposures. Any decision will always have some disadvantages and should be justified in the sense that it should do more good than harm. In these circumstances, as stated in Para. 207 of *Publication 103* (ICRP, 2007a), the principle of justification is primarily applied in industries involving NORM when deciding whether to implement a protection strategy for radiation exposures.

(48) As such, justification falls under the ethical values of beneficence, which means promoting or doing good, and non-maleficence, which means avoiding causation of harm (ICRP, 2018), in order to reach the overall goal of societies to contribute to the well-being of individuals, and the quality of living together with the preservation of biodiversity and sustainable development.

(49) As explained in Para. 208 of *Publication 103* (ICRP, 2007a), responsibility for judging justification usually falls on governments or other national authorities to ensure that an overall benefit results, in the broadest sense, to society and thus not necessarily to each individual. However, input to the justification decision may include many aspects that could be informed by the industry involving NORM, workers, the public, and organisations other than the government or national authority. As such, justification decisions regarding radiological protection strategies could benefit from a stakeholder involvement process. In this context, radiological protection considerations will serve as one input to the broader decision-making process.

(50) The need for a protection strategy controlling NORM exposure is better understood after radiological characterisation of the exposure situation, and taking into account health, economic, societal, and ethical considerations. Since many industries involving NORM already exist, the Commission suggests the establishment at national level of a list of industries involving NORM for which a radiological risk assessment should be undertaken in order to determine if a protection strategy is justified. The level of control may then be determined through implementation of the optimisation principle. If an ongoing industrial process involving

NORM, not previously identified on a national list, appears to be of concern, the justification of a protection strategy may be addressed on a case-by-case basis with involvement of the relevant stakeholders.

(51) For industries involving NORM on the national list, when a new process using NORM is to be implemented, the principle of justification should be applied in the same way as for ongoing processes (i.e. primarily when deciding whether to implement a protection strategy for radiation exposures). Industrial processes will usually produce significant economic and social benefits, and the radiological risks involved are unlikely to result in a decision that the NORM process, as a whole, would need to be considered unjustified. Exceptions can be dealt with on a case-by-case basis.

3.3. Optimisation of protection

(52) When a decision has been taken to implement a protection strategy, the principle of optimisation of protection becomes the driving principle to select the most effective actions for protecting the exposed public, workers, and the environment. As far as human protection is concerned, it is defined by the Commission as the process to keep the magnitude of individual doses, the number of people exposed, and the likelihood of incurring exposures as low as reasonably achievable, guided by appropriate individual dose criteria, taking into account economic and societal factors. The impact to the environment should also be kept as low as reasonably achievable. This means that the level of protection should be the best possible under the prevailing circumstances, adopting a prudent and reasonable attitude (ICRP, 2018).

(53) To avoid serious inequity in the individual dose distribution of humans, in line with the ethical value of justice (ICRP, 2018), the Commission recommends using individual dose criteria in the optimisation process (ICRP, 2007a, Para. 232). In addition to reduction of the magnitude of individual exposures, reduction of the number of exposed individuals should also be considered. The collective effective dose has been and remains a key parameter for optimisation of protection for workers in comparing the exposures predicted from different options for protection strategies.

(54) The optimisation process should consider protection of the environment. The aim is to avoid deleterious effects on non-human species. Such an approach should be commensurate with the overall level of risk, and compatible with common standards of environmental protection, notably the optimisation of discharges in the environment. As is the case for human exposure, NORM processes may pose environmental risks from other constituents, and radiological aspects have to be considered in an all-hazard approach. In practice, the radiological impact should be included in the environmental impact assessment and monitored as necessary. The approach already developed by the Commission (ICRP, 2008, 2014a), through a set of RAPs and numerical values for DCRLs, is useful guidance when assessing possible deleterious radiation effects on non-human species, their diversity,

communities, and ecosystems in general. The results contribute to decisions on the most appropriate option for controlling the source.

(55) In the case of industries involving NORM, the optimisation process is implemented in generally the same way as for other industries. However, because of the prevailing circumstances, and notably because radiological protection should be integrated in a broader protection strategy in which the radiological risk is not necessarily dominant, the options to reduce doses may be more limited and/or may require different resources. Such challenges suggest the need for flexibility in implementation of the optimisation process and application of regulatory structures.

(56) The involvement of relevant stakeholders early in the optimisation process will contribute to selecting the best option for protection, taking into account the characteristics of the actual exposure situation, and thus potentially making protection more effective and efficient.

3.3.1. Dose criteria

(57) The Commission recommends the use of reference levels as dose criteria in existing exposure situations. The reference level represents the value of dose used to guide and drive the optimisation process. Selection of the reference level should consider the actual individual dose distribution, with the objective of identifying those exposures that warrant specific attention. Reference levels are guides for selecting amongst protective options in the optimisation process in order to maintain individual doses as low as reasonably achievable, taking into account economic and societal factors, and thus prevent and reduce inequities in dose distribution. Reference levels are also benchmarks against which the results of protective actions can be judged to determine if protection is reasonably optimised and effective.

(58) As far as the protection of non-human species is concerned, the Commission recommends the use of DCRLs. DCRLs can be considered as a band of dose rates within which there is likely to be some chance of deleterious effects of ionising radiation occurring to individuals of that type of RAP (derived from a knowledge of defined expected biological effects for that type of organism). When considered together with other relevant information, the DCRL can be used as a point of reference to optimise the level of effort expended on environmental protection, dependent upon the overall management objectives and the exposure situation (ICRP, 2008). The way to achieve protection of the environment using DCRLs is described in Section 4.3.

(59) For the protection of humans in existing exposure situations, the Commission recommends setting reference levels typically within the 1–20 mSv year^{-1} band as presented in Table 5 of *Publication 103* (ICRP, 2007a), with the possibility that the most appropriate reference level to guide optimisation of protection could be $<$1 mSv year^{-1}. The 1–20 mSv year^{-1} band presupposes that the sources or pathways can generally be controlled, and individuals receive direct benefits from activities associated with the exposure situation, but not necessarily from the exposure itself. However, selection of the reference level for a particular exposure situation should be

made based upon the characteristics of the circumstances (ICRP, 2007a, Para. 234), considering the individual dose distribution, with the objective of identifying those exposures that warrant specific attention and meaningfully contribute to the optimisation process.

(60) Industries involving NORM generally give rise to low or moderate levels of individual exposure, and the appropriate reference level can, in most cases, be less than a few mSv per year. The selected reference level should be meaningful for protection purposes, not a generic value which would not help to identify individuals for whom some further consideration might be needed. Thus, according to the characteristics of the exposure situation, notably the actual and potential exposure pathways, the individual dose distribution, and the prospect for optimisation, an appropriate reference level can, in most cases, be less, perhaps well less, than a few mSv annual effective dose. If, in rare instances with larger individual doses in the dose distribution, a reference level could be selected above a few mSv, the Commission would expect that the level would rarely need to exceed 10 mSv annual effective dose. The reference level applies to the dose added to the natural background.

(61) Section 4 contains specific bands of reference levels that are recommended for the protection of NORM workers and the public, respectively. They are consistent with the approach recommended in *Publication 103* (ICRP, 2007a) and with the general approach to selection of a reference level as described above.

(62) The Commission recognises that some authorities have specified dose limits for some industries involving NORM, in addition to industries where NORM is processed for its radioactive, fissile, or fertile properties. This may be particularly suitable in circumstances when the source is well characterised and controlled, and there is ongoing potential for significant levels of exposure. However, specifying a limit for regulatory purposes does not imply that the complete framework for management of planned exposure situations has to be applied. The Commission recommends that an optimised (graded) approach is applied to NORM industries so that efforts and resources expended on protection are commensurate with the radiological hazards and risks, and that this should be taken into account when a decision is taken on control of the exposure. This means that the burden imposed by the regulatory system should be balanced with the outcomes to be achieved. Requirements that do not contribute meaningfully to achieving radiological and non-radiological protection should be avoided.

3.3.2. The optimisation process

(63) Optimisation of protection of human health and the environment in industries involving NORM is implemented through a process that involves: (a) assessment of the exposure situation; (b) identification of possible protective options to maintain or reduce exposure to as low as reasonably achievable, taking into account economic and societal factors; (c) selection and implementation of the most appropriate protective options under the prevailing circumstances; and (d) regular review of the

exposure situation to evaluate if there is a need for corrective actions, or if new opportunities for improving protection have emerged.

(64) In this iterative process, the Commission considers that the search for equity in the distribution of individual exposures is an important aspect (ICRP, 2006). It should be noted that in industries involving NORM, the distribution of individual doses for both workers and members of the public may be very wide. Protective efforts should focus on individuals in the higher dose tail of the distributions (i.e. the most exposed individuals) in order to determine if efforts are reasonable to reduce their exposures, while simultaneously trying to reasonably reduce the exposure of the whole exposed population.

(65) Decision-making for control of industries involving NORM should be open and transparent. Stakeholders should be involved as necessary, including workers, the community, and others as appropriate. Their concerns and ideas should be listened to and taken into account. A transparent system for decision-making will allow controversial issues to be properly addressed and resolved, although not necessarily with full agreement from all parties.

(66) The inclusion of natural or man-made radiation in a work environment highlights the need to foster development of an appropriate radiological protection culture within the organisation and community, enabling each individual to make well-informed choices and behave wisely in situations involving potential or actual exposure to ionising radiation (ICRP, 2006). It is a matter closely tied to the ethical concept of dignity (ICRP, 2018).

(67) Detailed advice from the Commission on how to apply the optimisation principle in practice has been provided previously (ICRP, 1983, 1990, 1991, 2006) and remains valid.

4. IMPLEMENTATION OF THE SYSTEM OF RADIOLOGICAL PROTECTION TO INDUSTRIAL PROCESSES INVOLVING NORM

4.1. Protection of workers

4.1.1. General considerations

(68) Typical industries involving NORM process a wide range of raw materials and activity concentrations. Depending upon the circumstances, it may not be necessary to consider controls directly applicable to a particular individual in order to properly control exposures. This does not mean that protection is not warranted, but that the control is exercised on the workplace and the conditions of work rather than on the worker her/himself. It is not easy to define criteria applicable in all situations. Thus, a graded approach for the protection of workers is recommended.

(69) The main exposure pathways for work with NORM are:

- external exposure (mostly due to gamma radiation, but occasionally beta-radiation exposure to the lens of the eye and the skin may need to be considered); and
- internal exposure from inhalation dust and, to a much lesser extent, ingestion of radioactive dust, as well as exposures due to radon gas and its progeny, which can occur above ground or underground (e.g. the build up of radon gas in underground workplaces), and sometimes thoron emanating from NORM. In practice, radon emanating from such materials is often indistinguishable from that already present (e.g. from the ground).

(70) The Commission considers that radon and thoron in the workplace, irrespective of the source, should be managed as a single source [i.e. as described in *Publication 126* (ICRP, 2014b)]. An integrated approach for protection against radon exposure in all buildings is recommended, whatever their purpose and the status of their occupants. The strategy of protection in buildings, implemented through a national action plan, should be based on application of the optimisation principle using a reference level translated for practical reasons to concentrations in air to facilitate implementation. The Commission recommends that national authorities should set a derived reference level that is as low as reasonably achievable in the range of 100–300 Bq m^{-3}, taking the prevailing economic and societal circumstances into account. The corresponding effective dose depends on a number of factors such as breathing rates [see *Publication 137* (ICRP, 2017b)]. As described in *Publication 126*, if radon mitigation actions cannot reduce levels to less than the reference level, the exposure will need to be considered as part of the occupational exposure.

(71) It is important to note that workers in industries involving NORM are exposed to radiation and also to other hazards. The radiological risk is often not the dominant hazard, and may historically not even have been a consideration.

In such a context, there is often a lack of radiological protection awareness or a culture supporting such protection. However, such industries do have experience and expertise in the management of occupational health and safety, and there is an opportunity to build a radiological protection culture in an integrated fashion. In many cases, actions to reduce workplace hazards, such as airborne dust, will also restrict radiation exposures, and an integrated approach to worker protection is recommended.

(72) Protection of workers in industries involving NORM should be based on a graded approach to control radiation exposures, according to the annual effective dose (due to the activities involving NORM) that is likely to be received and the scope for dose reduction that may be achievable through optimisation.

(73) In practice, a graded approach can be realised through selection of suitable dose reference levels, selection of the requisites (i.e. appropriate protective actions), and integrated implementation of these requisites. Practical implementation of this approach will also help to determine whether or not the workers should be considered as occupationally exposed to radiation.

(74) This approach can serve as the basis for creating a common understanding between regulatory authorities and other stakeholders – such as operators, workers and their representatives, and health, safety, and environmental professionals – of the radiological aspects of the various processes involved and the ways in which these aspects can be addressed reasonably and effectively.

4.1.2. Selection of the dose reference level for workers

(75) Since the industries involving NORM are so diverse, there is no unique numerical value which is appropriate as a reference level for all of them. According to the characteristics of the exposure situation, notably the actual and potential exposure pathways, the individual dose distribution, and the prospect for optimisation, the appropriate reference level can be selected based on the 1–20-mSv band recommended by the Commission, noting that the selection could be:

- of the order of a few mSv per year, or below, for most cases; and
- above a few mSv, but very rarely exceeding 10 mSv year^{-1}, when necessary because of the circumstances involved.

(76) Considering the current information about the distribution of doses to workers in many industries involving NORM, the selection of a reference level above 10 mSv year^{-1} would not be necessary in terms of radiological protection.

(77) As indicated above, these doses exclude exposures from radon or thoron. In *Publication 126* (ICRP, 2014a), the Commission recommends that national authorities should set derived reference levels for radon and thoron that are as low as reasonably achievable in the range of 100–300 Bq m^{-3}, taking the prevailing economic and societal circumstances into account. When concentrations still exceed the reference level following application of radon prevention and mitigation

measures, it may be necessary, within a graded approach, to undertake additional assessments of exposure in terms of dose; in such a case, a reference level of the order of 10 mSv should be used.

(78) In most situations, the residual doses are not expected to exceed the reference level, particularly after the effective implementation of protective measures. The reference level remains useful to allow judgement on appropriate functioning of the programme, and to indicate if modifications are needed.

4.1.3. Selection and implementation of requisites

(79) When considering measures to optimise exposures to NORM workers, the starting point should always be the existing industrial safety and hygiene controls (i.e. for non-radiological hazards in the workplace). Experience shows that a well-managed, safety-focused workplace will already have done much to reduce radiation exposures from NORM, and may, in fact, be adequate for radiological protection without further addition. Where additional radiological protection controls are considered necessary, these should be integrated into the wider safety strategy as far as practicable.

(80) The strategy for protection of workers as defined in conventions from the International Labour Organisation (Conventions 167 and 176) comprises three main steps:

- eliminate the hazard and consequently the risk (e.g. by replacing hazardous substances by harmless or less hazardous substances wherever possible);
- minimise the risk (e.g. by technical measures applied to the plant, machinery, equipment, or process); and
- in so far as the risk remains, undertake other effective measures related to the workers themselves (e.g. use of PPE).

(81) The same scheme is relevant for the protection of workers in industries involving NORM. Control of the workplace and the conditions of work are to eliminate or minimise the risk, while individual control is required when adequate protection has not already been achieved. Moving from control of the workplace to individual control needs careful consideration as this is costly, and the preference would be to have sufficient control of the workplace so that individual control is not needed (apart from, sometimes, simple PPE). The requisites related to the workplace and the conditions of work are described below.

4.1.3.1. Characterisation of the situation

(82) This characterisation, determining who is exposed, when, where, and how, is an important starting point for the protection of workers. It includes characterisation of the source, with the aim of identifying the distribution of NORM radionuclides and their activity concentrations throughout the industrial process,

including mode of exposure, chemical and physical characteristics of particulates, NORM distribution, and activity concentrations at all stages of the industrial process in both operational and maintenance conditions. Feed materials, intermediates, residues, wastes (including contamination of the plant), and discharges to the environment should be considered, as well as radon and thoron.

(83) Characterising the source will help to identify the main exposure pathways to workers, the public, and the environment. In terms of exposure of workers, the next step is to characterise exposed groups or individuals, and make an initial assessment of the annual doses (effective doses arising from external exposure and through inhalation) received from work with NORM.

(84) Characterisation of the exposure situation may, of course, vary in detail according to the prevailing circumstances involved. In practice, external gamma radiation and internal exposures from radioactive dust inhalation are the two exposure pathways of interest. In particular, inhalation of radon and thoron has to be considered, although it is recommended that this should be addressed separately. When considering the likely annual radiation exposure of workers in different industries involving NORM, it is important that these are based on realistic estimates (i.e. taking into account actual external radiation and airborne contamination levels in the workplace and actual working patterns and procedures). When estimating radiation exposures, the effect of existing occupational health and safety provisions should be taken into account (e.g. industrial hygiene, industrial safety, and workplace controls on airborne dust).

(85) It is important that this characterisation stage is fully documented in order to provide a sound basis for any future decision-making.

(86) The characterisation will form the basis for justification of the protection strategy, notably the need for specific requisites for radiological protection purposes, as well as for scaling of the optimisation process.

(87) The initial characterisation should be subject to periodic review. The detail and frequency of this periodic review should be commensurate with the level of risk. When feedstocks, ores, production practices, or other factors that can affect dose are expected to change significantly, a new characterisation should be undertaken.

4.1.3.2. Obtaining expert radiological protection advice

(88) Such expertise is normally required from the beginning (i.e. to assist with characterisation of the exposure situation). Typically, industries involving NORM have operated for many years before the issue of natural radioactivity has been addressed. As a result, there is often a complete lack of knowledge about radioactivity and radiological protection. Consequently, the first step should be to seek expert advice on this issue, even where industries involving NORM already have their own technical support in a wide range of other areas. Such specific expertise can be provided either internally or by external consultants. Such radiological protection

expertise should be sought by both operating management and by the national authorities where no specialised expertise exists. The need for advice from a radiological protection expert may be temporary (e.g. where it can be shown from an initial review and assessment that exposures are very low) or may be required on an ongoing basis.

4.1.3.3. Initial actions to prevent or reduce the hazard

(89) This corresponds to the first step of the International Labour Organization approach. At the initial stage, it is useful to consider if there are any ways in which the hazards from NORM can be either eliminated from the process or substantially reduced. Examples include the selection of alternative feed materials (i.e. with much lower concentrations of NORM), or changes to the process designed to prevent the accidental accumulation or concentration of radionuclides. Whilst recognising that this is likely not practical or possible, particularly in long-standing industries involving NORM, it should nevertheless be given some consideration.

4.1.3.4. Delineation of areas

(90) The delineation of areas is a well-established element of the control strategy in planned exposure situations. However, it is also part of a wider industrial health and safety strategy [i.e. to identify areas where additional safety measures (e.g. working procedures, ventilation requirements, use of PPE, limitation of access) are required]. Different areas may be delineated in the same installation because they correspond to a given type of management (e.g. radon area vs other areas). To be effective, area delineation requires warning signs and, in some cases, formal restrictions on access. The same approach is appropriate for industries involving NORM. Worker right-to-know protocols may determine the type of signage needed. The concept may already be in place in some industries, as there would, for example, be warnings and controls for dust.

4.1.3.5. Engineered controls

(91) As mentioned above, the characteristics of NORM are such that scenarios involving high doses from accidental exposures do not generally exist. Thus, the traditional engineering controls to prevent such exposures are not required. Instead, measures to restrict long-term exposures from NORM are the more important consideration. These start with the design and layout of the facility, and then specific measures to control dust (e.g. containment and ventilation). Industries involving NORM, such as mineral processing plants, may be very dusty, and a dust control strategy and programme should already be in place in such facilities. Improvements to containment and ventilation systems should be

considered holistically (i.e. in terms of their overall effect on radioactive and other materials).

(92) Specific engineering measures to restrict external radiation exposures (i.e. shielding) may be required; for example, local shielding around pipes and vessels containing NORM at very high activity concentrations may be considered. More commonly, however, protection is provided through adjustments to working patterns and, in some cases, relocation of materials, plant, or persons (distance).

4.1.3.6. Working procedures

(93) These procedures, such as limiting time of exposure, can be very effective in restricting both internal and external doses, even where exposures are already low. Often, all that is required is observance of good industrial hygiene and simple safe working procedures, supported by an appropriate amount of training (see below) and supervision.

(94) The requisites listed above, complemented by at least a general information programme for workers (see below), may be sufficient for the protection of workers in most industries involving NORM. However, they can be complemented, as necessary, by requisites related to the individuals.

4.1.3.7. Information, instruction, and training

(95) The information and training provided to workers should be proportionate to the radiation risk and the precautions that need to be taken. There is a basic need to share information and generally raise awareness about NORM within the workplace. In particular, information should be provided to pregnant and breastfeeding workers. NORM workers are key stakeholders in this process, and the principles of open communication and engagement should be applied at an early stage. Where special precautions to restrict exposures to radiation are required, the relevant workers should receive specific training to understand the nature of the radiological risks and the importance of protective actions, and practical instructions on how to implement these actions.

4.1.3.8. Personal protective equipment

(96) This includes protective clothing and respiratory protective equipment (e.g. dust masks), and these are already widely used in NORM workplaces to protect against other hazards. PPE should be selected with due consideration of the hazards involved. The equipment should not only provide adequate protection but also be convenient and comfortable to use. The effectiveness of any existing PPE should be assessed before determining whether improved or additional PPE for radiological

protection purposes is required. Engineered controls are the favoured option, with working procedures and, finally, protective respiratory equipment being considered only where further engineering controls are not effective or practicable. Consideration should also be given to the possibility of an increase in exposure caused by the additional constraints of PPE.

4.1.3.9. Dose assessment

(97) An assessment of the exposure of workers is required as part of the initial characterisation described above. It is envisaged that this will be based on workplace measurements and other information (e.g. about the process and working practices), rather than individual dosimetry. In practice, although the level of dose may not be the only criterion, where worker doses are estimated to be higher than a few mSv per year, an ongoing programme of dose assessment should be implemented, according to a graded approach. Where doses are above a few mSv per year, it is expected that they may be estimated on the basis of workplace measurements. Individual dose assessment (e.g. through the use of personal dosimeters) may be useful as a means of providing information to help optimise exposures, but is not expected to be undertaken on a routine basis.

(98) Where doses are well above a few mSv per year, individual dose assessments should be undertaken. For external radiation, this should be done with personal dosimeters (passive or electronic). Assessment of internal exposures from dust inhalation is much more challenging; however, in very dusty NORM workplaces, there may already be a dust monitoring programme which can be adapted to provide estimates of radiation dose. If not, and if internal doses are high, arrangements with a suitable internal dosimetry service will need to be considered. It should be noted, however, that such exposures are unlikely to be considered optimised, and that suitable protective actions should be more than capable of reducing internal exposures.

(99) As far as radon and thoron are concerned, exposure should be assessed, but not necessarily in terms of dose, as long as the concentration can be controlled. When radon or thoron dose assessment is relevant, it may be performed through collective or individual monitoring, or inferred from monitoring of the workplace (ICRP, 2014b).

4.1.3.10. Dose recording

(100) Both workplace and individual data related to the estimation and assessment of worker doses should be recorded and kept for sufficient time. The recording may be carried out in different ways according to the situation. For instance, it could be by keeping track of ambient exposure in a given place of work and of people who frequented it, so as to be able to assess the doses of a given worker retrospectively if

necessary. It could also be carried out by registering individual doses in the dedicated sheet in the medical record of each concerned worker.

4.1.3.11. Health surveillance

(101) In some industries involving NORM, there is already a health surveillance programme for non-radiological reasons. It is considered unlikely that health surveillance specifically for radiological protection purposes will be required, except in a very few cases where annual doses well above a few mSv are received repeatedly. If this is the case, it is expected that existing provisions for the health surveillance of workers occupationally exposed to radiation will be used, and will be sufficient.

(102) Most of these requisites need to be implemented only to the extent necessary to achieve acceptable protection. The modalities for implementing the requisites should also be adapted to the circumstances. Workers are likely to be considered as occupationally exposed when, despite all reasonable efforts to reduce exposure, elevated individual doses persist and application of special working procedures is needed to perform the job. In the case of radon exposure, *Publication 126* (ICRP, 2014b) recommends that workers may also be considered as occupationally exposed in some workplaces identified in a national list of activities, or facilities in which workers are inevitably and substantially exposed to radon, and this exposure is more intimately and obviously related to their work activities.

(103) In these cases, education and training, individual radiation dose monitoring and recording, or health surveillance for radiological protection purposes may all need to be implemented as described in *Publication 75* (ICRP, 1997).

4.2. Protection of the public

(104) The general approach to protection of the public should start with characterisation of the exposure situation (who is exposed, when, where, and how), including analysis of exposure pathways and dose assessments. This characterisation forms the basis for justification of a protection strategy. Next, the optimisation process should be implemented, including selection of a reference level, selection and implementation of protective actions, involvement of stakeholders in the decision-making process, and provision of long-term monitoring of the situation if necessary.

(105) This process should be implemented in a reasonable way, keeping in mind the ethical values of beneficence/non-maleficence, prudence, justice, and dignity. In more complex situations, working with stakeholders to identify their underlying interests for each ethical value can be very useful in working towards an acceptable and sustainable solution.

(106) The reference level for protection of the public should be selected of the order of a few mSv per year, or below, to meaningfully guide the process of optimisation of protection. In some cases of public exposure for industries involving

NORM, a reference level <1 mSv year^{-1} may, in fact, be the most appropriate, taking into account the distribution of doses that exists. Protection of the public should be addressed as a whole (i.e. taking into account the different pathways). In a given situation, the pathways need to be considered with respect to NORM discharges, wastes, residues, and possible legacy sites. In practice, the most exposed individuals to each pathway belong to different groups so the reference level can generally be applied to any given pathway. The reuse and recycling of NORM residues may be the starting point of a new NORM process.

(107) Public exposure to radon or thoron arising from industries involving NORM is mainly due to the reuse of residues (e.g. in building materials). The corresponding recommendations set in *Publication 126* (ICRP, 2014b) are included in Section 4.2.4.

4.2.1. Discharges from industries involving NORM

(108) Liquid and gaseous radioactive and/or non-radioactive effluents may be discharged deliberately from the normal operation of industries involving NORM. Radionuclides may also change their physicochemical form [e.g. some may react with particles in the stream of liquid or gaseous effluents (aerosols)]. In certain cases, such as oil and gas extraction, the phosphate processing industry, and the combustion of coal, NORM discharges have been an issue for the protection of both people and the environment. Therefore, effluents should be properly controlled, taking into account the radiological and non-radiological impacts and, if necessary, restricted in order to protect the public and the environment.

(109) A comprehensive site-specific control of discharges should, from a radiological protection point of view, include the following steps:

- radiological characterisation of discharges;
- identification of potential exposure pathways, taking into account the environmental distribution of radionuclides in space and time, as well as radionuclide mobility under ambient conditions;
- dose assessments and risk estimation;
- justification of measures to control discharges;
- selection of a reference level; and
- selection and implementation of measures within a protection strategy through an optimisation process (as low as reasonably achievable).

(110) The protection strategy should include preventive actions aimed at eliminating or reducing the quantity and concentration of discharges, as well as mitigation actions aimed at reducing the impact of discharges in terms of public and environmental exposures. The optimisation process and the involvement of stakeholders are case specific and depend, in practice, on the operational characteristics of the NORM facility, discharge processes, radioactivity levels and estimated risk, public groups involved, as well as societal and political aspects and public awareness. In practice, optimisation can be complex due to the fact that some processes such as effluent

treatment may lead to the production of further waste in which there are increased concentrations of radionuclides, or else produce an increase in the overall volume of waste produced.

(111) Attention should also be paid to the issue of drinking water, environmental impact (see below), current and future land use in the area, and the possible presence of several facilities in the same area.

(112) The use of reference levels translated into a measurable quantity (e.g. in terms of total activity and/or activity concentration) may be appropriate for industries involving NORM.

4.2.2. Wastes

(113) Wastes, both liquid and solid, are materials with no further planned use. Industries involving NORM can produce wastes containing both radioactive and non-radioactive pollutants; both should be managed consistently. Globally, industries involving NORM produce wastes ranging from small volumes with high concentrations of radionuclides to large volumes with low concentrations of radionuclides.

(114) Wastes should be characterised in order to determine the proper methods for disposal. Waste treatment should be considered and performed as relevant in the optimisation process, although concentration of wastes to high levels can pose challenges. The issue of wastes should be considered from their generation to final disposal when starting or designing a new project ('from cradle to grave').

(115) The method of disposal of NORM wastes should be proportionate to the type and level of hazard, taking into account all types of pollutants (radioactive and non-radioactive). Depending on the level of radioactivity and volume of wastes, a graded approach should apply. Some wastes could be treated as industrial or hazardous wastes and disposed of accordingly in near-surface landfills. The disposal of wastes with higher concentrations of radionuclides should be consistent with the management of radioactive wastes.

4.2.3. Residues

(116) Residues are materials which can be recycled and/or reused. They mainly come from upstream of the NORM cycle (exploration, extraction of material), and the activity concentration in the residues may be enhanced significantly compared with the raw material. Like wastes, they should be characterised and properly stored before potential reuse. There are economic and ecological arguments for finding a use for NORM residues. By-products and residues of a given industry involving NORM can be used as feedstock by other industries involving NORM, as landfill (if there are no chemical hazards or pathways to groundwater), and/or in commodities (e.g. building materials). Using residues as feedstock may be the starting point of a new NORM process. Recycling or reuse helps to reduce waste volumes. However, in some cases, it could result in exposure of workers, the public, and the

environment. Residues that are stockpiled for any length of time should be properly managed to prevent environmental contamination.

(117) The implementation of a protection strategy should be considered for reuse or recycling of NORM residues. The assessment should take account of various elements such as the level of exposure, pollution of the environment, alternatives, future of the products, and societal acceptance. In rare cases, based on this assessment, the new process may not be justified and the residues may need to be treated as wastes.

(118) When a protection strategy is justified, optimisation should be considered, recognising that the scope for dose reduction may be limited.

4.2.4. Building materials

(119) Building materials may contain natural radionuclides originating from raw materials (e.g. extracted from quarries) or residues from industries involving NORM or a mixture of materials, some of which are naturally radioactive (e.g. concrete). They can cause public exposures by direct external gamma radiation and by releasing radon and thoron into indoor air. Occupational exposures in the manufacture and handling of building materials are usually low, but they should be managed in a graded approach as in any other industry involving NORM.

(120) The use of building materials containing NORM may be considered as one pathway of public exposure to NORM, such as discharges, wastes, and residues, to which, as explained above, the reference level for public exposure of the order of a few mSv per year or below can apply. It should be expressed as effective dose caused by external gamma radiation to members of the public. A reference level of this order should also ensure that any radon exhalation from ^{226}Ra in building materials is generally unlikely to be the cause for the reference levels set for indoor radon concentration to be exceeded. In most countries, building materials are of minor importance for radon exposure, but there can be special cases where that source cannot be neglected. The exhalation of thoron is expected to be of less concern.

(121) Lists of building materials, raw materials, and residues of concern may be found in various publications (EURATOM, 2013; IAEA, 2015). There are also different methodologies for screening building materials of concern and for assessing the dose caused by building materials (EC, 1999b; IAEA, 2005; EURATOM, 2013). However, to provide information on the actual exposure caused by a building material, more elaborate methods need to be used in order to consider the actual concentrations and locations of a certain building material in a building (EC, 1999b; EURATOM, 2013; IAEA, 2015).

(122) A protection strategy should be established with the aim of promoting building materials that do not exceed the reference level. The strategy may encompass measures such as providing information on the levels of exposure caused by different building materials, labelling of materials, suggesting the use of materials with low radioactive concentrations, or limiting the use of certain materials causing significant exposures. In keeping with the ethical value of beneficence/non-maleficence,

it is important to ensure that the measures envisaged are actually reasonable and feasible before deciding on them.

(123) The national radon action plan established as recommended in *Publication 126* (ICRP, 2014b) should include radon and thoron exposure from building materials as relevant. Actions aimed at prevention and mitigation of such exposure in buildings, irrespective of its origin, are provided in *Publication 126*, and should be undertaken as necessary to reduce radon exposures.

(124) Special attention should be paid to processes where residues with exceptionally high activity concentrations are incorporated into building materials. They should not be implemented for the purpose of intentional diluting or for bypassing more stringent requirements on the appropriate management of such residues. This applies irrespective of whether the reference level for building materials might be exceeded.

(125) There may be a need to apply a similar approach for other construction materials, such as those used for foundations of houses; surfaces of yards, playgrounds, streets, and roads; bridges; and other similar structures. Dose assessments and separate derived activity concentration indexes may need to be considered.

4.2.5. Legacy sites

(126) Industries involving NORM account for many current legacy sites with radioactive contamination. NORM legacy sites have been identified more frequently with the rising awareness of industries involving NORM and related radiological protection issues. This situation shows that radiological protection is sometimes not considered sufficiently when facilities are shut down and dismantled. Technologies and methods already exist and should be implemented during the operation of the process involving NORM in order to prevent legacy sites.

(127) The issue of legacy sites is the scope of a future ICRP publication; therefore, the present publication provides only a few general considerations. The assignment of responsibility or liability for maintenance and remediation of old legacy sites may be an issue due to the elapsed time and often lost information. Sites with no known responsible party are often called ‘orphan sites’. New legacy sites should be avoided through proper dismantling of the industries involving NORM and durable administrative control if necessary.

(128) Justification of the remediation of legacy sites is not only driven by radiological protection considerations. As in active industries involving NORM, other hazards such as heavy metals may also be present. The reference level should be in the lower range of the band 1–20 mSv year^{-1}. The reference level is not the endpoint of the remediation. The endpoint should be an optimised level of dose below the reference level, determined on a case-by-case basis taking into account the prevailing circumstances (including the situation pre-disturbance), future use of the site (when it can be predicted), and possible conditions (or restrictions) of use.

(129) Implementation of the optimisation principle is often a challenge, for example because it is occasionally difficult to make a distinction between NORM

contamination and natural background radioactivity. The challenge may also be due to a lack of societal acceptance of the legacy and even of its management. The involvement of stakeholders in the decision process is of great importance for the management of legacy sites.

(130) Workers involved in the remediation process may need to be specifically trained for working with radiation. In that case, they should be considered as occupationally exposed.

(131) If common workers or members of the public are participating in the remediation (in their home or in places open to the public), relevant information and recommendations should be communicated to them as well as protective equipment, such as respiratory protection, as relevant.

4.3. Protection of the environment

(132) Large quantities of NORM may be present in the environment in the form of mixed material together with other contaminants. Over time, different geochemical and physical processes in the environment disturb the NORM radionuclide equilibrium. It is well known that mechanisms such as selective dispersion, leaching and transfer, fractionation, bioaccumulation, and reaction with other contaminants can result in changes in environmental impact over time. In this type of environmental exposure, it can be difficult to use a simple approach for risk assessments to evaluate the possible risk and effects for non-human species.

(133) The optimisation process should address protection of the environment (i.e. protection of non-human species) and not only the prevention of exposure of humans through environmental pathways (ICRP, 2007a). Mechanisms to control releases of effluents, in particular, can be informed by prediction of dose for non-human biota. The selected controls may, or may not, be specifically driven by radiological protection for non-human species, but the relative contribution for different options is useful information. However, the information on elevated NORM activity concentration in a given environmental compartment does not necessarily mean induction of effects in non-human species, and the assessment of impact should consider a variety of factors beyond estimated dose.

(134) Over the last few decades, considerable international and national efforts have been made to develop an approach for radiological protection of the environment. Raising awareness about radioactivity in industrial activities has become important at both national and international levels. Industries involving NORM have generally been following common standards to protect the environment from pollutants other than radioactivity.

(135) The Commission recommends an integrated approach which should encompass:

- all stressors/factors of concern (i.e. radiological and non-radiological); and

- human health effects due to environmental exposure of humans and ecological effects due to environmental exposure of non-human species and their assemblage (i.e. from populations of species to communities and ecosystems).

(136) The main issue is harmonious and consistent demonstration of the protection of people and the environment with a balanced, well-justified and integrated approach. Generally, this approach can be implemented in a graded way, as recommended in any environmental impact assessment, by starting with a very simple conservative assessment (screening stage making use of generic input data under the assumption of cautious exposure scenario) and then, if needed, by increasing the complexity and realism of the assessment as necessary (e.g. by using site-specific data and more detailed and realistic exposure scenarios) until a clear and defensible conclusion is reached (IAEA, 2018).

(137) Any industry may foresee an added value of implementing a generic case for a group of similar facilities and/or practices as a common screening integrated approach. In this case, the demonstration should be convincing and based on relevant source term and exposure scenarios, with the aim of driving an easy-to-implement screening out of sites with trivial concerns for human or ecological health on one hand and radiological or non-radiological on the other hand.

(138) Protective actions could also be developed either by complementing such generic cases or by being considered as site specific where cultural, sociological, and economical aspects also have to be taken into account to drive the decision-making process involving the stakeholders.

(139) In the case of complex situations, the radiological characterisation of NORM released in the environment may be performed by the analysis of radionuclides with respect to their physical and chemical forms and activity concentrations both to the source and in the environmental media of concern (air, water, sediment, soil). To be able to assess exposure of non-human species, it may be further relevant to identify mobility of radionuclides, spatial and temporal variation, environmental pathways to plants and animals, and bioavailability. An approach with RAPs and DCRLs has been developed (ICRP, 2008, 2014b). Dosimetry models to calculate specific exposure doses from chosen radionuclides and for ecosystems and organisms of concern have been available for site-specific use. A degree of precaution may be considered necessary because of the importance of the site or habitat, or the importance of the actual species present or likely to be present. It is important to note that, in many cases, other constituents which present hazards to plants and animals will also be present. The Commission re-emphasises its recommendation that an all-hazard approach should be undertaken.

(140) The environmental impact assessment can be used as a basis for the justification of actions aimed at the protection of both human and non-human species, because decisions on restricting discharges will impact all types of exposure. The involvement of stakeholders is recommended. The long-term preservation of the environment is a global concern of society, to which application of the ethical values of radiological protection can usefully contribute.

(141) When dealing with NORM discharges in the environment, special requisites concerning radionuclides, time interval for analysis, samples to be analysed, organisms of concern, record keeping, and monitoring plan should be specified as necessary. Long-term environmental monitoring should be performed regularly to check if the protection criteria are still being met.

5. CONCLUSIONS

(142) NORM in industrial processes may be an issue from a radiological protection point of view. The industries are diverse, do not correspond to a sector in themselves, and are often large industries of economic importance. The way to address radiological protection in industries involving NORM has been a matter of discussion for decades. It is a matter of justice and equity, which are ethical values of the system of radiological protection, to consider radiological aspects as well as other industrial and chemical hazards. Doses from industries involving NORM are variable, but they can be comparable to, or greater than, those arising from other human activities already applying the system of radiological protection. However, in industries involving NORM, it is very unlikely that doses will reach a level leading to tissue reaction.

(143) Industries involving NORM are generally subject to authorisation, although, in many cases, not for radiological protection purposes, and these industries are familiar with risk management frameworks for the protection of workers, the public, and the environment. They should generally be able to apply the criteria and requisites set for radiological protection purposes. Experience shows that it is easier to develop a multihazards approach starting from conventional health and safety standards rather than from the system of radiological protection. In that context, the Commission recommends a realistic and pragmatic attitude.

(144) Industrial processes using NORM, although diverse, have specificities that have to be taken into account in a protection strategy. Often, such industries have been ongoing for a long time, with concern about radiological protection being a relatively recent addition. They are multihazards situations and, in most cases, the radiological risk is not dominant. While industries involving NORM have experience in risk management, they often have limited awareness of radiological protection; this can and should be developed.

(145) Industries involving NORM may need to be controlled, and the system of protection, including the principles of justification and optimisation of protection, as well as the corresponding dose criteria and requisites, can be applied. For adaptation to the features of industries involving NORM, the Commission recommends considering, as a starting point, the protection strategies already implemented by these industries to manage the hazards they are facing, and then estimating, after characterisation, the need for further action for protection against radiation. Such an integrated approach can be graded with proper balance between the different hazards, adopting a reasonable and prudent attitude and taking into account economic and societal considerations. Involvement of the relevant stakeholders in the decision process is crucial.

REFERENCES

EC, 1999a. Establishment of Reference Levels for Regulatory Control of Workplaces where Materials are Processed which Contain Enhanced Levels of Naturally Occurring Radionuclides. Radiation Protection 107. European Commission, Brussels.

EC, 1999b. Radiological Protection Principles Concerning the Natural Radioactivity of Building Materials. Radiation Protection 112. European Commission, Brussels.

EURATOM, 2013. Council Directive 2013/59/EURATOM of 5 December 2013 laying down basic safety standards for protection against the dangers arising from exposure to ionising radiation and repealing Directives 89/618/Euratom, 90/641/Euratom, 96/29/Euratom, 97/43/Euratom and 2003/122/Euratom. Off. J. Eur. Union. 13: 1–73.

IAEA, 1996. International Basic Safety Standards for the Protection against Ionizing Radiation and for the Safety of Radiation Sources. Safety Series No 115. International Atomic Energy Agency, Vienna.

IAEA, 2005. Environmental and Source Monitoring for Purposes of Radiation Protection. Safety Guide No. RS-G-1.8. International Atomic Energy Agency, Vienna.

IAEA, 2006. Assessing the Need for Radiation Protection Measures in Works Involving Minerals and Raw Materials. Safety Reports Series No. 49. International Atomic Energy Agency, Vienna.

IAEA, 2010. Proceedings of Naturally Occurring Radioactive Material Symposium (NORM VI). STI/PUB/1497, Marrakech, Morocco, March 2010. International Atomic Energy Agency, Vienna.

IAEA, 2014. Radiation Protection and Safety of Radiation Sources: International Basic Safety Standards. General Safety Requirements No. GSR Part 3. International Atomic Energy Agency, Vienna.

IAEA, 2015. Protection of the Public against Exposure Indoors due to Radon and Other Natural Sources of Radiation. Specific Safety Guide No. SSG-32. International Atomic Energy Agency, Vienna.

IAEA, 2018. Protective Radiological Environmental Impact Assessment for Facilities and Activities. General Safety Guide No. GSG-10. International Atomic Energy Agency, Vienna.

ICRP, 1977. Recommendations of the ICRP. ICRP Publication 26. Ann. ICRP 1(3).

ICRP, 1983. Cost–benefit analysis in the optimization of radiation protection. ICRP Publication 37. Ann. ICRP 10(2/3).

ICRP, 1984. Principles for limiting exposure of the public to natural sources of radiation. ICRP Publication 39. Ann. ICRP 14(1).

ICRP, 1990. Optimization and decision making in radiological protection. ICRP Publication 55. Ann. ICRP 20(1).

ICRP, 1991. 1990 Recommendations of the International Commission on Radiological Protection. ICRP Publication 60. Ann. ICRP 21(1–3).

ICRP, 1993. Protection against radon-222 at home and at work. ICRP Publication 65. Ann. ICRP 23(2).

ICRP, 1997. General principles for the radiation protection of workers. ICRP Publication 75. Ann. ICRP 27(1).

ICRP, 1999. Protection of the public in situations of prolonged radiation exposure. ICRP Publication 82. Ann. ICRP 29(1/2).

ICRP, 2006. The optimisation of radiological protection: broadening the process. ICRP Publication 101, Part 2. Ann. ICRP 36(3).

ICRP, 2007a. The 2007 Recommendations of the International Commission on Radiological Protection. ICRP Publication 103. Ann. ICRP 37(2–4).
ICRP, 2007b. Scope of radiological protection control measures. ICRP Publication 104. Ann. ICRP 37(5).
ICRP, 2008. Environmental protection – the concept and use of reference animals and plants. ICRP Publication 108. Ann. ICRP 38(4–6).
ICRP, 2009a. Application of the Commission's recommendations for the protection of people in emergency exposure situations. ICRP Publication 109. Ann ICRP 39(1).
ICRP, 2009b. Application of the Commission's recommendations to the protection of people living in long-term contaminated areas after a nuclear accident or a radiation emergency. ICRP Publication 111. Ann. ICRP 39(3).
ICRP, 2009c. Environmental protection: transfer parameters for reference animals and plants. ICRP Publication 114. Ann. ICRP 39(6).
ICRP, 2010. Lung cancer risk from radon and progeny, and statement on radon. ICRP Publication 115. Ann. ICRP 40(1).
ICRP, 2014a. Protection of the environment under different exposure situations. ICRP Publication 124. Ann. ICRP 43(1).
ICRP, 2014b. Radiological protection against radon exposure. ICRP Publication 126. Ann. ICRP 43(3).
ICRP, 2016. Radiological protection from cosmic radiation in aviation. ICRP Publication 132. Ann. ICRP 45(1).
ICRP, 2017a. Dose coefficients for nonhuman biota environmentally exposed to radiation. ICRP Publication 136. Ann. ICRP 46(2).
ICRP, 2017b. Occupational intakes of radionuclides: Part 3. ICRP Publication 137. Ann. ICRP 46(3/4).
ICRP, 2018. Ethical foundations of the system of radiological protection. ICRP Publication 138. Ann. ICRP 47(1).
ILO, 1988. C167 – Safety and Health in Construction Convention. Entry into force: 11 January 1991. Adoption: Geneva, 75th ILC session (20 June 1988). International Labour Organization, Geneva.
ILO, 1995. C176 – Safety and Health in Mines Convention. Entry into force: 5 June 1998. Adoption: Geneva, 82nd ILC session (22 June 1995). International Labour Organization, Geneva.
Miller, H.T., Bruce, E.D., Cook, L.M., 1991. Management of Occupational and Environmental Exposure to Naturally Occurring Radioactive Materials (NORM). 1991 SPE Annual Technical Conference and Exhibition, Pt. 2, Production Operations and Engineering. Society of Petroleum Engineers of AIME, 6–9 October 1991, Richardson, TX, USA, pp. 627–636.
Monicard, R., Dumas, H., 1952. Radioactivité des Roches Sédimentaires, du Pétrole Brut et des Eaux de Gisements. Institut Français du Pétrole, Paris, pp. 96–102.
Schmidt, A.P., 2000. Naturally Occurring Radioactive Materials in the Gas and Oil Industry. Origin, Transport and Deposition of Stable Lead and ^{210}Pb from Dutch Gas Reservoirs. Department of Geochemistry, Utrecht University, Utrecht.
UNSCEAR, 1977. Report to the General Assembly. United Nations Scientific Committee on the Effects of Atomic Radiation, New York.

UNSCEAR, 1982. Report to the General Assembly, Annexe C. United Nations Scientific Committee on the Effects of Atomic Radiation, New York.

UNSCEAR, 2008. Report to the General Assembly, Annexe B. United Nations Scientific Committee on the Effects of Atomic Radiation, New York.

UNSCEAR, 2016. Report to the General Assembly, Annexe B. United Nations Scientific Committee on the Effects of Atomic Radiation, New York.

ANNEX A. ACTIVITIES GIVING RISE TO NORM EXPOSURES

(A1) The main activities giving rise to NORM exposure are as follows.

A.1. Extraction of rare earth elements

(A2) The most important sources of rare earth elements are monazite (Ce, La, Nd, Th)PO_4 and bastnaesite. The crystal structure of monazite accepts uranium and thorium, and is the most common radioactive mineral on Earth. Activity concentration ranges from 5000 to 350,000 Bq kg^{-1} of ^{232}Th, and from 10,000 to 50,000 Bq kg^{-1} of ^{238}U (UNSCEAR, 2008). During the extraction process to obtain rare earth elements (by mechanical or chemical means), inhalation of dust and external gamma radiation to workers may occur. Furthermore, effluents, residues, and wastes from the extraction process contain thorium, radium, and uranium at concentrations higher than in the feedstock (EC, 1999a). Waste in the form of mill tailings can be used for landfill material or may need specific management.

A.2. Production and use of metallic thorium and its compounds

(A3) Thorium in oxide form occurs in many minerals, notably monazite. It can be extracted by concentrating minerals and decomposing them with acid to obtain thorium salts; this is the raw material for the production of thorium in metallic form. Thorium is used in a number of materials, usually as an additive [e.g. thoriated tungsten isolated welding electrodes, that usually contain 100,000 Bq kg^{-1} of ^{232}Th and ^{228}Th (EC, 1999a)] or alloy (e.g. magnesium thorium used in jet engines; activity of approximately 70,000 Bq kg^{-1}), and as thorium nitrate in the manufacture of gas mantles. Small quantities of thorium can be found in many products: glass, airport runway lights, lamp starters, etc. Producing material containing thorium can give rise to external gamma exposure and internal exposure through the inhalation of dust. The process also generates solid wastes and effluents that may need to be monitored and controlled.

A.3. Mining and processing of ores (other than uranium)

(A4) According to the International Labour Organization, mining is an extensive industry that accounts for approximately 1% of the world workforce (i.e. approximately 30 million workers, including some 12 million in coal mining). The main source of exposure in mining operations is radon; however, exposure due to long-term radionuclides through gamma external exposure and the inhalation and ingestion of mineral dusts can be important in certain situations.

(A5) The processing of ores may also be affected by the use of NORM, and exposure situations for workers differ considerably with respect to the type of industry, conditions in workplaces, radionuclides involved and their physical and chemical forms, etc. The natural radionuclides involved in extractive industries end up in the products and/or in the effluents and/or wastes. Sediment discharges in waste water into the environment have been measured with activity up to 55,000 Bq kg^{-1} of ^{226}Ra and 15,000 Bq kg^{-1} of ^{228}Ra (IAEA, 2003).

A.4. Extraction of oil and gas

(A6) The water contained in oil and gas geological formations contains ^{228}Ra, ^{226}Ra, and ^{224}Ra dissolved from the reservoir rock, together with their decay progenies. When this water is brought to the surface with the oil and gas, changes in temperature and pressure can lead to the precipitation of radium-rich sulphate and calcium carbonate scales on the inner walls of production equipment (pipes, valves, pumps, etc.). Depending on the age of the scale, significant amounts of ^{210}Pb and ^{228}Th may grow in with their respective radioactive parents (IAEA, 2006). In any case, the activity concentrations in scale are difficult to predict, and the activity concentration of ^{226}Ra has been reported to range from <1000 to ~1,000,000 Bq kg^{-1} (EC, 1999a). Radium isotopes and their progeny can also appear in sludges in separators and skimmer tanks [more details can be found in Table 5 of IAEA (2003)]. The main radiological protection issues associated with scale are external gamma exposure of workers, especially where scale is deposited, and internal exposure of staff removing scale during maintenance and decommissioning. Figures related to activity concentrations in oil, gas, scale, and sludge are given in Table A.1 (IAEA, 2003, 2011).

(A7) Operators may try to prevent deposition of scale through the application of chemical scale inhibitors in the water. As a result, the radium isotopes will pass through the production system and be released with the produced water. In the

Table A.1. Range of concentrations of radionuclides in oil, gas, and by-products.

	Crude oil (Bq kg^{-1})	Natural gas (Bq m^{-3})	Produced water (Bq l^{-1})	Hard scale (Bq kg^{-1})	Sludge (Bq kg^{-1})
^{238}U	0.0001–10		0.0003–0.1	1–500	5–10
^{226}Ra	0.1–40		0.002–1200	100–15,000,000	5–800,000
^{210}Po	0–10	0.002–0.08		20–1500	4–160,000
^{210}Pb		0.005–0.02	0.05–190	20–75,000	100–1,300,000
^{222}Rn	3–17	5–200,000			
^{232}Th	0.3–2		0.0003–0.001	1–2	2–10
^{228}Ra	3–17		0.3–180	50–2,800,000	500–50,000
^{224}Ra			0.5–40		

same way, the new technique of 'fracking' (hydraulic fracturing) for gas production also releases NORM in drill cuttings and water. For example, the US Geological Survey shows a median activity concentration for produced water of 200 Bq L^{-1} (Rowan et al., 2011).

A.5. Manufacture of titanium dioxide

(A8) Titanium can be extracted from ilmenite (which contains monazite as an impurity) and rutile which may contain elevated levels of both ^{232}Th and ^{238}U. Radiological exposure from titanium dioxide production varies with the type and source of ore and the process. Ore concentration activity of ^{238}U and ^{232}Th ranges from 7 to 9000 Bq kg^{-1} (EC, 1999a). The separation process could give rise to radiological hazards from dust inhalation and external gamma radiation emanating from large stockpiles of material. Precipitate containing isotopes of radium may occur during the process and be found in the waste [at activity concentrations up to 1,600,000 Bq kg^{-1} (IAEA, 2006)].

A.6. The phosphate processing industry

(A9) Phosphate rock is the starting material for the production of all phosphate products and is the main source of phosphorus for fertilisers. The radionuclide content of the ore varies greatly depending on its origin (IAEA, 2003), and is generally <3000 Bq kg^{-1} of uranium. Phosphate processing can be divided into the mining and milling of phosphate ore (there is no significant enhancement of activity concentration during this phase, but exposure through inhalation and external exposure may occur) and the manufacturing of phosphate products by wet or thermal processes.

(A10) Most phosphate rock is treated with sulphuric acid to produce phosphoric acid (wet process). The phosphoric acid can be combined with ammonia to make ammonium phosphate which is the basis of mixed fertiliser. The production of phosphoric acid generates large quantities of phosphogypsum, and evidence suggests that radium isotopes are more readily retained in phosphogypsum (EC, 1999a). Phosphogypsum is also used as a building material and in agriculture. Environmental protection issues (regarding radiological impact and toxicity) may arise from the disposal of phosphogypsum in stockpiles or by discharge into surface water bodies.

(A11) Furthermore, radium scale and sediment can be formed inside equipment during wet processes, and radium activity concentrations in scale vary from values similar to those in the original ore up to 1000 times greater (IAEA, 2006), leading to possible exposure by external gamma radiation and/or inhalation of dust during maintenance and decommissioning.

(A12) In the thermal process, phosphate is crushed and mixed with silica and coke to be burnt in a furnace at 1500°C. At this temperature, phosphorus vapour is produced and can be condensed and removed as liquid or solid. The elemental phosphorus can be used for the production of high-purity phosphoric acid and other phosphorus products. During this process, volatile radionuclides such as ^{210}Pb and ^{210}Po are also produced and become concentrated in the precipitator [typical concentrations of 50,000–500,000 Bq kg^{-1} (EC, 1999a)], while thorium and uranium are retained in the slag (activity concentrations range between 1 and 3000 Bq kg^{-1}). Dust and slag may present NORM exposure to workers and the public when used as construction material in cement.

A.7. The zircon and zirconia industries

(A13) Zircon (or zirconium silicate) is recovered from beach sand. The sand is pre-processed in very large quantities by gravimetric and electromagnetic sorting to separate the mineral sands. Exposure of workers to NORM may arise due to the inhalation of dust and external irradiation from the large amount of material. When chemical processing of zircon is used, effluents may contain NORM. A very large range of activity concentrations are reported for zirconium silicate, from 200 to 74,000 Bq kg^{-1} of ^{238}U and from 400 to 40,000 Bq kg^{-1} of ^{232}Th (EC, 1999a; IAEA, 2012). Most zircon sand is used as an opacifier in fine ceramics, enamels, glazes, and sanitary ware. Zircon sand can also be manufactured as a refractory component by mixing the sand with alumina and sodium carbonate, and smelting the mixture. ^{210}Pb and ^{210}Po are volatilised and end up in the fume collection system [up to 200,000 Bq kg^{-1} of ^{210}Pb and 600,000 Bq kg^{-1} of ^{210}Po (IAEA, 2006)].

A.8. Production of metal

(A14) Largely depending on the origin of the metal ore, the extraction of many metals may give rise to exposure to NORM because smelting and refining at high temperatures may volatilise ^{210}Pb and ^{210}Po from ore that can lead to exposure by inhalation during the process, and later when these radionuclides have been precipitated and concentrated [up to 200,000 Bq kg^{-1} (IAEA, 2006, 2013)]. Non-volatile radionuclides may be concentrated in the slag (range from <1000 to $>10,000$ Bq kg^{-1}). Such exposures could occur in the production of tin, copper, iron, steel, aluminium, niobium/tantalum, bismuth, etc.

A.9. Extraction and combustion of coal

(A15) Most fossil fuels, notably coal, contain uranium and thorium and their decay products, as well as ^{40}K. The activity concentrations are generally not elevated and depend on the region of origin and its geology [examples of figures are given on p. 184 of UNSCEAR (2016)]. However, UNSCEAR estimated that occupational exposure due to coal mining was 23,000 man.Sv for the 2002–2003 period, and that the annual average effective dose for Chinese coal miners (90% of the work-force) was 2.75 mSv $year^{-1}$. Due to the amount of material, the quantities of radio-nuclides involved are noteworthy. For example, over 8000 million tons of coal were extracted in 2014 (according to British Petroleum Statistical Review of World Energy), and by considering the lower values of 4 ppm of uranium and 10 ppm of thorium, 32,000 tons of uranium and 80,000 tons of thorium can be considered as being extracted as well.

(A16) The combustion of coal fuel to produce heat and electricity generates fly ash and the heavier bottom ash or slag. The concentration of radionuclides in bottom ash and slag tends to be higher than in coal (around 10 times), but generally does not exceed 5000 Bq kg^{-1} (IAEA, 2006); the range of radionuclide activities in ashes are presented in Table A.2 (UNSCEAR, 1982). Volatile materials such as lead and polonium can be released to the atmosphere or, in modern power stations, retained, and can accumulate in fly ash as well as the inner surface of the burner (^{210}Po activity concentration >100,000 Bq kg^{-1} in deposited scale has been reported). Gas desul-phurisation results in additional sludge and gypsum. The use of coal combustion residues (ash, gypsum) in cement or concrete is a worldwide practice.

Table A.2. Ranges of radionuclide activities in coal ash and slag.

	Potassium (Bq kg^{-1})	Thorium series (Bq kg^{-1})	Uranium series (Bq kg^{-1})
Bottom ash (slag)	240–1200	44–560	48–3900
Fly ash (collected)	260–1500	30–300	30–2000
Fly ash (escaping)	260	100–160	20–5500

A.10. Water treatment

(A17) Treatment of underground water is a common practice to remove salts and other contaminants. Various processes may be used, such as filters or ion exchange resins. Radionuclides of natural origin present in underground water may accumu-late in water treatment wastes (filter sludge). The activity concentrations in such wastes are generally moderate but can reach 10,000 Bq kg^{-1} (IAEA, 2006).

A.11. Building materials

(A18) The use of some building materials may lead to elevated indoor radiation levels when they contain elevated levels of radionuclides, particularly ^{226}Ra, ^{232}Th,

Table A.3. Examples of activity concentrations (in Bq kg^{-1}) for some building materials.

Material	^{226}Ra	^{232}Th	^{40}K
Concrete	1–250	1–190	5–1570
Aerated concrete	11,000	1–220	180–1600
Clay bricks	1–200	1–200	60–2000
Sand-lime bricks and sandstone	18,000	11,000	5–700
Natural gypsum	<1–70	<1–100	7–280
Granite	100	80	1200
Lithoid tuff	130	120	1500
Pumice stone	130	130	1100
Cement	7–180	7–240	24–850
Tiles	30–200	20–200	160–1410
Phosphogypsum	4–700	19,000	25–120
Blast furnace slag stone and cement	30–120	30–220	–

and ^{40}K. The building material may be of natural origin or contain materials derived from industrial processes such as those listed above. Values for activity concentrations (in Bq kg^{-1}) in some building materials are given in Table A.3 (UNSCEAR, 1982; IAEA, 2003).

(A19) Activity concentration guidelines for the use of NORM in building materials have been developed in Europe through the use of an activity concentration index, considering ^{226}Ra, ^{232}Th, and ^{40}K activity in the material (EC, 1999b; EURATOM, 2013).

A.12. Legacy sites

(A20) There are also several sites with residues from former installations around the world. Most of these sites are contaminated with natural radionuclides from former industries involving NORM. In some cases, these sites have been identified and successfully remediated. However, it is almost certain that a significant number of contaminated sites from former industries involving NORM have yet to be identified.

(A21) Industries involving NORM process a wide range of raw materials with large variation in activity concentrations, producing a variety of products, by-products, and wastes which have an even larger variation in activity concentrations. These industries may or may not be of concern depending on the activity concentrations in the raw materials handled, processes adopted, uses for final products, reuse and recycling of residues, and disposal of wastes.

A.13. References

EC, 1999a. Establishment of Reference Levels for Regulatory Control of Workplaces where Materials are Processed which Contain Enhanced Levels of Naturally Occurring Radionuclides. Radiation Protection 107. European Commission, Brussels.

EC, 1999b. Radiological Protection Principles Concerning the Natural Radioactivity of Building Materials. Radiation Protection 112. European Commission, Brussels.

EURATOM, 2013. Council Directive 2013/59/EURATOM of 5 December 2013 laying down basic safety standards for protection against the dangers arising from exposure to ionising radiation and repealing Directives 89/618/Euratom, 90/641/Euratom, 96/29/Euratom, 97/43/Euratom and 2003/122/Euratom. Off. J. Eur. Union. OJ L 13, 17.1.2014, p. 1–73.

IAEA, 2003. Extent of Environmental Contamination by Naturally Occurring Radioactive Material (NORM) and Technological Options for Mitigation. Technical Reports Series No. 419. International Atomic Energy Agency, Vienna.

IAEA, 2006. Assessing the Need for Radiation Protection Measures in Works Involving Minerals and Raw Materials. Safety Reports Series No. 49. International Atomic Energy Agency, Vienna.

IAEA, 2011. Radiation Protection and NORM Residue Management in the Production of Rare Earths from Thorium Containing Minerals. Safety Reports Series No. 68. International Atomic Energy Agency, Vienna.

IAEA, 2012. Radiation Protection and NORM Residue Management in the Titanium Dioxide and Related Industries. Safety Reports Series No. 76. International Atomic Energy Agency, Vienna.

IAEA, 2013. Radiation Protection and Management of NORM Residues in the Phosphate Industry. Safety Reports Series No. 78. International Atomic Energy Agency, Vienna.

Rowan, E.L., Engle, M.A., Kirby, C.S., Kraemer, T.F., 2011. Radium Content of Oil- and Gas-field Produced Waters in the Northern Appalachian Basin (USA). Summary and Discussion of Data. U.S. Geological Survey Scientific Investigations Report 2011–5135. US Geological Survey, Reston, VA. Available at: http://pubs.usgs.gov/sir/2011/5135/ (last accessed 12 September 2019).

UNSCEAR, 1982. Report to the General Assembly, Annexe C. United Nations Scientific Committee on the Effects of Atomic Radiation, New York.

UNSCEAR, 2008. Report to the General Assembly, Annexe B. United Nations Scientific Committee on the Effects of Atomic Radiation, New York.

UNSCEAR, 2016. Report to the General Assembly, Annexe B. United Nations Scientific Committee on the Effects of Atomic Radiation, New York.

ACKNOWLEDGEMENTS

At its meeting in Berlin (Germany) in October 2007, the Main Commission of ICRP approved the formation of Task Group 76, reporting to Committee 4, to develop recommendations to cover the broad range of activities associated with the processing, manufacturing, use, and disposal of materials with enhanced levels of NORM. The publication should also clarify issues concerning the type of exposure situation, categories of exposure, and basic principles to be applied for the management of NORM.

ICRP thanks all those involved in the development of this publication for their hard work and dedication over many years.

Task Group 76 members *(2010–2013)*

P. Burns (Chair)	M. Markkanen	Å. Wiklund*
A. Canoba	S. Romanov	D. Wymer*
A. Liland	L. Setlow	
G. Loriot		

*Corresponding members

Task Group 76 members *(2013–2019)*

J-F. Lecomte (Chair)	F. Liu	P.P. Haridasan (–2015)*
D. da Costa Lauria	M. Markkanen	H.B. Okyar (–2017)*
P. Egidi	P. Shaw (–2017)	S. Mundigl*
A. Liland		

*Corresponding members

Committee 4 critical reviewers

A. Canoba	T. Pather (2013–2017)	G. Hirth (2017–2021)

Main commission critical reviewers

C-M. Larsson	S. Romanov

Editorial members

C.H. Clement (Scientific Secretary and *Annals of the ICRP* Editor-in-Chief)
H. Fujita (Assistant Scientific Secretary and *Annals of the ICRP* Associate Editor) *(2018–)*
H. Ogino (Assistant Scientific Secretary and *Annals of the ICRP* Associate Editor) *(2016–2018)*

Committee 4 members during preparation of this publication

(*2009–2013*)

J. Lochard (Chair)
W. Weiss (Vice-Chair)
J-F. Lecomte (Secretary)
P. Burns
P. Carboneras
D.A. Cool
T. Homma
M. Kai
H. Liu
S. Liu
S. Magnusson
G. Massera
A. McGarry
K. Mrabit
S. Shinkarev
J. Simmonds
A. Tsela
W. Zeller

(*2013–2017*)

D.A. Cool (Chair)
K-W. Cho (Vice-Chair)
J-F. Lecomte (Secretary)
F. Bochud
M. Boyd
A. Canoba
M. Doruff
E. Gallego
T. Homma
M. Kai
S. Liu
A. McGarry
A. Nisbet
D. Oughton
T. Pather
S. Shinkarev
J. Takala

(*2017–2021*)

D.A. Cool (Chair)
K.A. Higley (Vice-Chair)
J-F. Lecomte (Secretary)
N. Ban
F. Bochud
M. Boyd
A. Canoba
D. Copplestone
E. Gallego
G. Hirth
T. Homma
C. Koch
Y. Mao
N. Martinez
A. Nisbet
T. Schneider
S. Shinkarev
J. Takala

Main Commission members at the time of approval of this publication

Chair: C. Cousins, *UK*
Vice-Chair: J. Lochard, *France*
Scientific Secretary: C.H. Clement, *Canada*; *sci.sec@icrp.org*[*]

K.E. Applegate, *USA*
S. Bouffler, *UK*
K.W. Cho, *Korea*
D.A. Cool, *USA*
J.D. Harrison, *UK*
M. Kai, *Japan*
C-M. Larsson, *Australia*
D. Laurier, *France*
S. Liu, *China*
S. Romanov, *Russia*
W. Rühm, *Germany*

Emeritus members
R.H. Clarke, *UK*
F.A. Mettler Jr, *USA*
R.J. Pentreath, *UK*
R.J. Preston, *USA*
C. Streffer, *Germany*
E. Vañó, *Spain*

[*]Although formally not a member since 1988, the Scientific Secretary is an integral part of the Main Commission.

ICRP and the members of Task Group 76 thank S. Andresz (CEPN) for his fruitful scientific assistance as Secretary of the Task Group, and L. Matta, J. Popic, and B. Wang for their helpful contributions to this publication through the ICRP consulting process.

Finally, thank you very much to all organisations and individuals who took the time to provide comments on the the draft of this publication during the consultation process.